# MAPPING THE NATION

## GIS for Good Governance

Esri Press
Redlands, California

Ask for Esri Press titles at your local bookstore or order by calling 800-447-9778, or shop online at esri.com/esripress. Outside the United States, contact your local Esri distributor or shop online at eurospanbookstore.com/esri.

Esri Press titles are distributed to the trade by the following:

*In North America:*
Ingram Publisher Services
Toll-free telephone: 800-648-3104
Toll-free fax: 800-838-1149
E-mail: customerservice@ingrampublisherservices.com

*In the United Kingdom, Europe, Middle East and Africa, Asia, and Australia:*
Eurospan Group
3 Henrietta Street
London WC2E 8LU
United Kingdom
Telephone: 44(0) 1767 604972
Fax: 44(0) 1767 601640
E-mail: eurospan@turpin-distribution.com

**All images courtesy of Esri except as noted.**

# Contents

# Socioeconomics | 33

## National Security | 69

## Credits | 84

# Foreword

When my wife, Laura, and I founded Esri as an urban planning consultancy in 1969, GIS heralded promise as a system of record and a data-sharing mechanism for all governance— not just urban planning. Conceptually speaking, GIS could simplify the organization, analysis, and expression of all data. Today, enabling technologies such as distributed computing have invigorated that concept, and now GIS is a primary tool for investigation, reporting, and decision-making for policy makers. The purpose of Esri's annual national map book is to highlight the exemplars of that fulfilled promise and connect it to the work you do.

This volume, *Mapping the Nation: GIS for Good Governance,* conveys the strong influence of mapping in developing more intelligent policy and propagating leaner, more efficient government. In the less than fifty years since GIS revolutionized urban systems, most federal agencies have now come to rely on cartography and geographic analysis to understand the socioeconomic, public safety, and environmental issues we face as a nation. The maps in this book show how GIS translates the complexity of these matters into stories we all intuitively understand.

I hope the tremendous work featured in this book opens your eyes to the limitless possibilities of GIS and inspires you to apply geography to meet your constituents' needs. By doing that, you'll continue innovating across all levels of government to build a greater nation.

Warm regards,

Jack Dangermond

# Introduction

When people's lives are affected by nationwide issues, time is critical; from the time it takes for elected officials to become aware of an issue at a community level, to the time it takes for policy makers to identify a solution and pass legislation on a national level.

The faster government can understand the nation's needs, the faster it can improve quality of life. That's what good governance is all about—taking informed action for the good of the people.

Intelligent mapping technology—known as geographic information system (GIS) software—is making it possible for decision-makers to quickly discern and resolve constituent needs by collaborating across all levels of government.

With GIS, policy makers are able to better share, access, visualize, and analyze information on a local, state, and national level to form policy that truly effects change.

Every community across the country is faced with its own unique challenges—from social and economic, to public safety and environmental concerns. So how does government identify issues that span local and state borders and require national attention?

Policy makers rely on data to paint the picture. They depend on agencies at all levels of government to supply the data they need to make informed decisions.

Traditionally, this information is delivered in reports or spreadsheets, but seeing data in these forms can only do so much. As the use of GIS increases in government, policy makers are becoming increasingly aware that visualizing big data on a map and analyzing it geographically is the fastest way to understand relationships and trends.

With GIS-derived maps and spatial analysis, elected officials are not only able to visualize and investigate their own polling data, they can also enrich it with authoritative content, demographic data, imagery, and maps from other agencies. Integrating this content and seeing it altogether on a map reveals insights fast—so elected officials can cut down on the time it takes to develop strategies for legislation.

To garner legislative support in Congress, members present maps that quickly demonstrate the problem across fellow legislators' districts. Being able to visualize the problem in a member's district or state often inspires action and support, in the same manner that led the initial sponsor to pursue the legislation.

When maps provide clear evidence that an issue must be addressed, policy makers across party and agency lines are able to come to a unified conclusion. And the application of GIS doesn't stop there. The White House and agencies implement and monitor policy using the technology. For many agencies, GIS is mission critical.

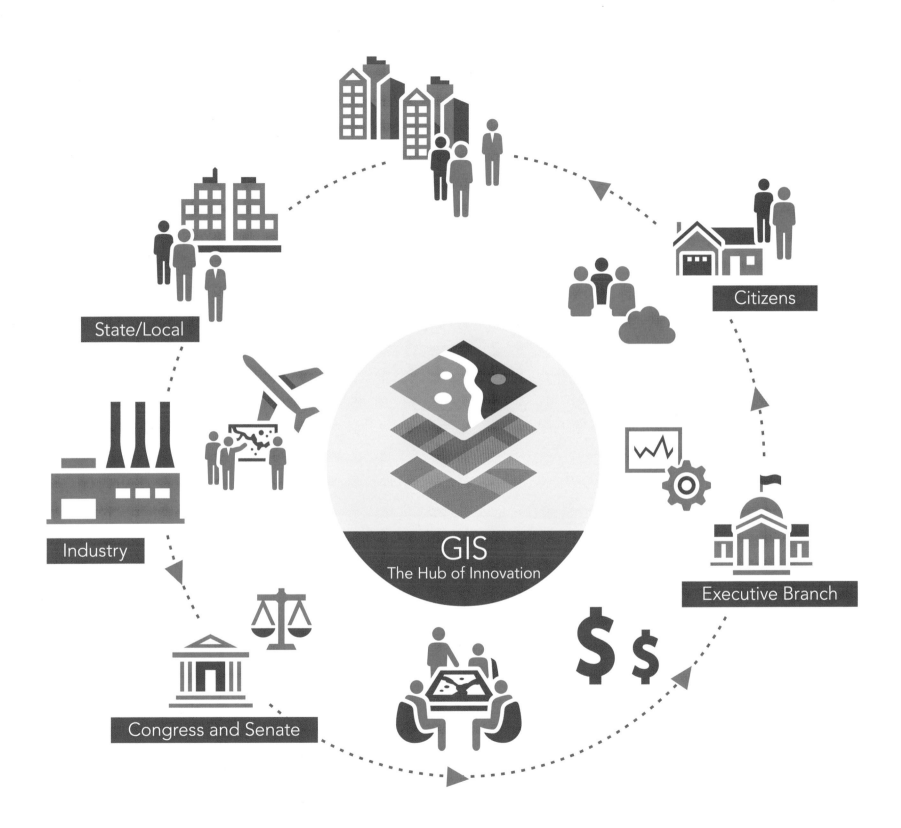

State/Local

Citizens

Industry

GIS
The Hub of Innovation

Executive Branch

Congress and Senate

$ $

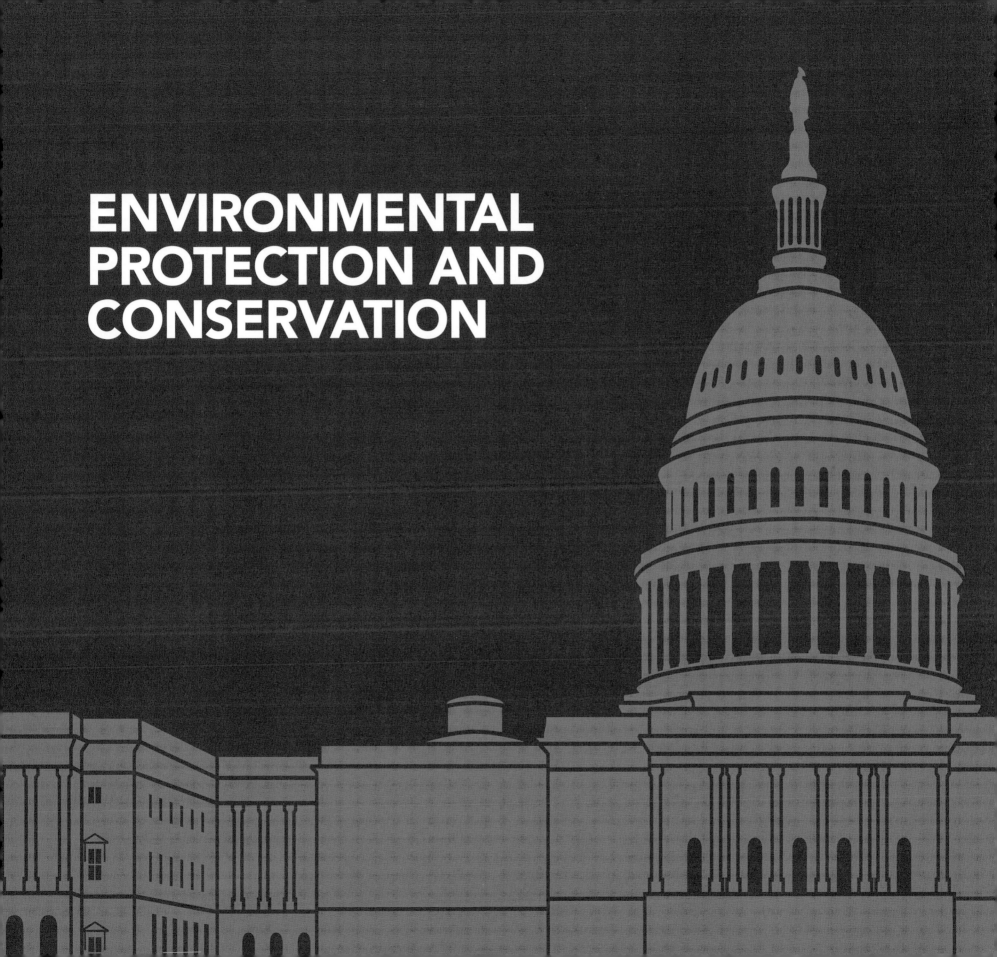

# ENVIRONMENTAL PROTECTION AND CONSERVATION

"As we continue our mission of protecting and preserving our natural and cultural heritage, the sign is clearly visible: Web GIS is moving us forward."

**Darcee Killpack**
GIS Coordinator, Intermountain Region
National Park Service

"Through web services and data standards and GIS processing tools and search tools, we are becoming much better at sharing our data. But now we have the opportunity to more fully tell the stories of our mission."

**Gary Latzke**
Wisconsin Internet Mapping Program Manager
US Geological Survey

# ENVIRONMENTAL PROTECTION AND CONSERVATION

Conservation and protection of our country's natural resources, wildlife, and landscapes has rippling benefits to the economy, quality of life, and environment. In today's technologically-driven world, organizations use GIS in all areas of their missions to make smarter decisions—from analyzing environmental problems and identifying solutions, to monitoring operations and educating the public of the benefits of sustainable development and conserving natural resources.

A key application of GIS among organizations—both government and nonprofit—is using maps to communicate and collaborate with elected officials and constituents. Maps make it easy for the public and policy makers to interact with data and easily understand what cultural and environmental resources are at risk, and how they can make a difference. Together, nonprofits and elected officials are using GIS to quickly analyze situations and identify smart solutions that support local, state, and national green infrastructure initiatives.

As conservation projects evolve from grassroots efforts to national priorities, nongovernmental organizations are also using GIS to deliver innovative reports that track and communicate the benefits of federal support. Take the Chesapeake Bay Program, for example. In 2016, the project commemorated its thirty-third year of collaboration among local, state, and federal organizations to protect and restore the Chesapeake Bay Watershed from the effects of pollution. Multiple organizations, including local high schools, used GIS to create story maps that demonstrate the ongoing environmental challenges and progress being made in the bay thanks to actions of the US Environmental Protection Agency and its state partners.

# Keep the Canyon Grand

Grand Canyon Trust

The Grand Canyon, one of the Seven Natural Wonders of the World, faces many challenges and obstacles regarding preservation of its land. The canyon was first home to the Havasupai, Hualapai, Hopi, Navajo, Paiute, Zuni, and other native people.  Enriched with mineral wealth including copper and uranium deposits, the Grand Canyon has been adversely affected by years of mining uranium on the South Rim. Other threats include inappropriate development and new wells that would stress the aquifer that feeds the canyon's precious seeps and springs. This story map conveys the history and beauty that lies within the Grand Canyon and points to resources on how to help protect the canyon.

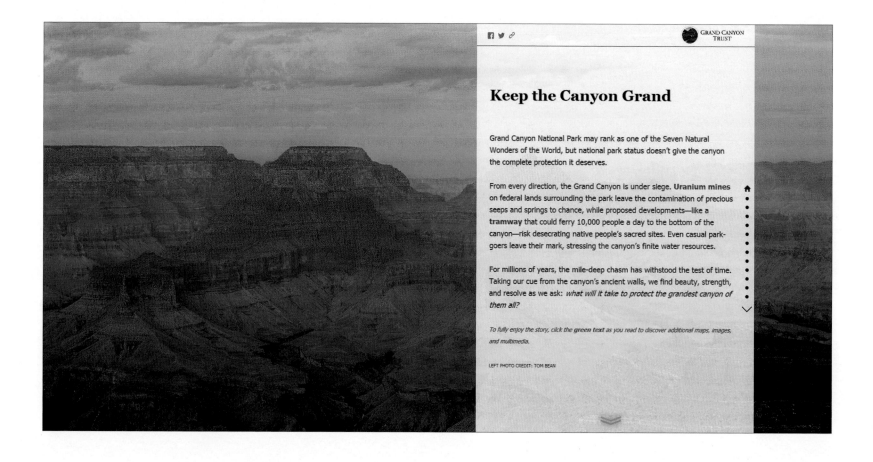

**Keep the Canyon Grand**

Grand Canyon National Park may rank as one of the Seven Natural Wonders of the World, but national park status doesn't give the canyon the complete protection it deserves.

From every direction, the Grand Canyon is under siege. **Uranium mines** on federal lands surrounding the park leave the contamination of precious seeps and springs to chance, while proposed developments—like a **tramway that could ferry 10,000 people a day** to the bottom of the canyon—risk desecrating native people's sacred sites. Even casual park-goers leave their mark, stressing the canyon's finite water resources.

For millions of years, the mile-deep chasm has withstood the test of time. Taking our cue from the canyon's ancient walls, we find beauty, strength, and resolve as we ask: *what will it take to protect the grandest canyon of them all?*

*To fully enjoy the story, click the green text as you read to discover additional maps, images, and multimedia.*

LEFT PHOTO CREDIT: TOM BEAN

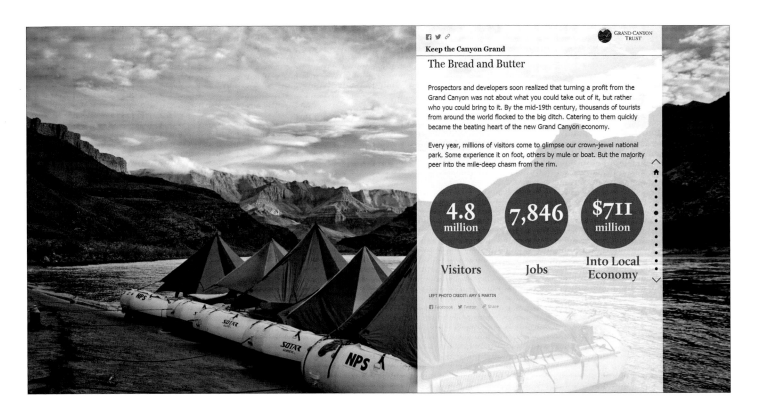

## The Bread and Butter

Prospectors and developers soon realized that turning a profit from the Grand Canyon was not about what you could take out of it, but rather who you could bring to it. By the mid-19th century, thousands of tourists from around the world flocked to the big ditch. Catering to them quickly became the beating heart of the new Grand Canyon economy.

Every year, millions of visitors come to glimpse our crown-jewel national park. Some experience it on foot, others by mule or boat. But the majority peer into the mile-deep chasm from the rim.

**4.8 million**
Visitors

**7,846**
Jobs

**$711 million**
Into Local Economy

LEFT PHOTO CREDIT: AMY S MARTIN

Facebook · Twitter · Share

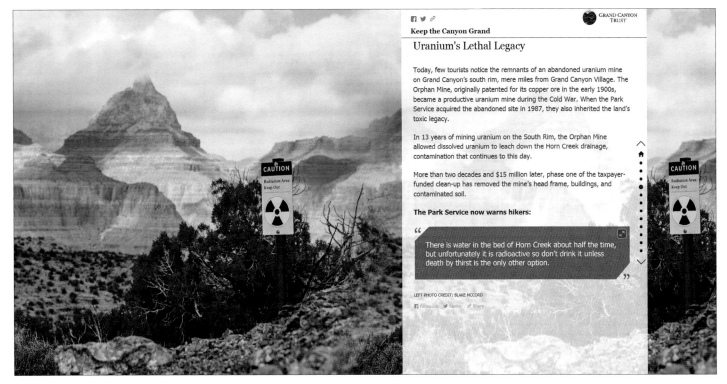

## Uranium's Lethal Legacy

Today, few tourists notice the remnants of an abandoned uranium mine on Grand Canyon's south rim, mere miles from Grand Canyon Village. The Orphan Mine, originally patented for its copper ore in the early 1900s, became a productive uranium mine during the Cold War. When the Park Service acquired the abandoned site in 1987, they also inherited the land's toxic legacy.

In 13 years of mining uranium on the South Rim, the Orphan Mine allowed dissolved uranium to leach down the Horn Creek drainage, contamination that continues to this day.

More than two decades and $15 million later, phase one of the taxpayer-funded clean-up has removed the mine's head frame, buildings, and contaminated soil.

**The Park Service now warns hikers:**

> There is water in the bed of Horn Creek about half the time, but unfortunately it is radioactive so don't drink it unless death by thirst is the only other option.

LEFT PHOTO CREDIT: BLAKE MCCORD

Facebook · Twitter · Share

# 100 Years of the National Park Service

National Park Service

August 2016 marked the centennial of the National Park Service (NPS). This story map is a chronological marking of significant events in the establishment and growth of America's unparalleled system of public parks, now totaling more than four hundred. The story begins eighty years before establishment of the NPS.

## January 2, 1893

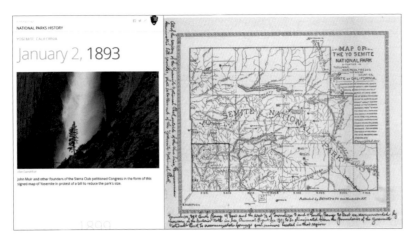

John Muir and other founders of the Sierra Club petitioned Congress in the form of this signed map of Yosemite in protest of a bill to reduce the park's size.

## March 1, 1903

Conservationist John Muir and President Theodore Roosevelt camped in Yosemite in 1903, and posed for a portrait at Glacier Point (click here to fly to Glacier Point). Roosevelt became a strong advocate for protection of Yosemite as a national park.

A story map

NATIONAL PARKS HISTORY

WASHINGTON DC

# August 25, 1916

President Woodrow Wilson signed the "Organic Act" into law, creating the National Park Service as part of the U.S. Department of the Interior. It consolidated management of 14 national parks, 21 national monuments, and the Hot Springs and Casa Grande Ruin reservations under its first director, Stephen Mather (below).

Right: a 1916 map highlights the national park system and the rail routes accessing them. See full-size version here.

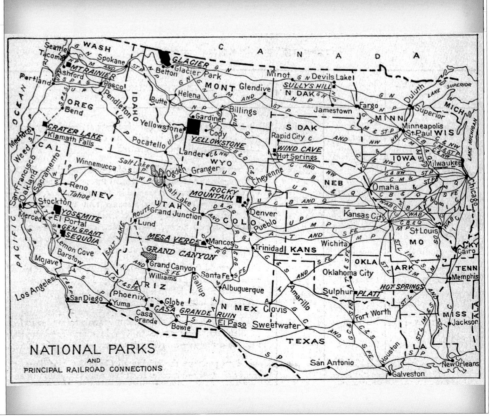

# Explore Virginia's National Parks

Office of US Senator Tim Kaine

Virginia Senator Tim Kaine uses his website to share his love of the outdoors with constituents and invite them to celebrate the National Park Service's 100th anniversary. A video message by Senator Kaine encourages constituents to find their parks and share their photos with him via social media. The page also includes the map of national parks in Virginia so that constituents have a place to get started. Each chapter within the map journal app represents one of Virginia's several national parks. Scrolling through the chapters brings viewers to the park's location on the map. The map also displays a pop-up window, such as this one over Shenandoah, which includes hyperlinks to appropriate hours and fees associated with the park.

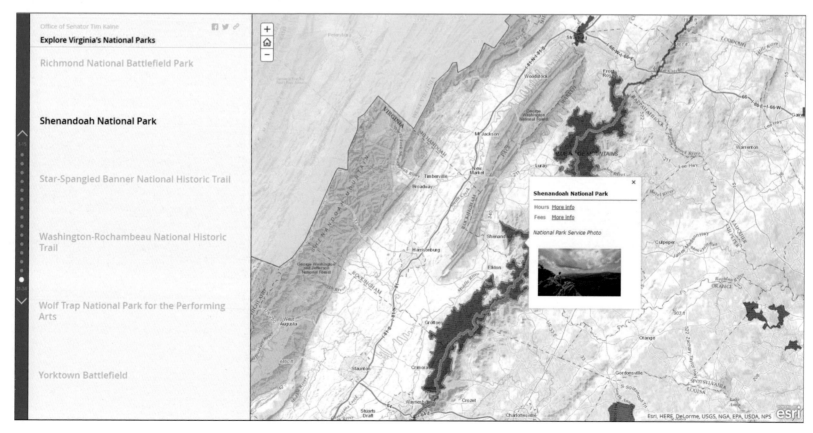

# Maps from Senate Analytical Mapping System: Arrowrock Dam Height Analysis, Boise River Streamflow Discharge Analysis

Office of US Senator James E. Risch

Members of Congress serve their constituents most effectively by keeping them informed and hearing their concerns. Senator James E. Risch of Idaho uses maps from the Senate Analytical Mapping System (SAMS) on his website to help accomplish this. Through maps, his constituents can visualize such important geographical information as locations of active forest fires within the state or places of congressionally funded projects. Utilizing SAMS also enables the senator to identify residents who share issues of concern according to their geographic concentration. Here are two examples of environmental issues in the senator's state:

### Arrowrock Dam Height Analysis

Upon the occurrence of an increased height for a dam, ArcGIS was used to display colored lines to indicate various water level elevations to visualize possible increases in reservoir storage.

### Boise River Streamflow Discharge Analysis

As Boise, Idaho's capital, experiences continued growth, watershed and streamflow discharge tools in ArcGIS help develop the best policies to sustain city expansion and support science policy for the next generation.

# Explore a Tapestry of World Ecosystems

US Geological Survey

The US Geological Survey (USGS) has published a new global ecosystems map of unprecedented detail to explore a tapestry of world ecosystems. This story map was produced by a team at the USGS Land Change Science Program. It is a mosaic of almost four thousand unique ecological areas called ecological land units (ELUs) based on four factors that are key in determining the makeup of ecosystems. Three of these—bioclimate, landforms, and rock type—are physical phenomena that drive the formation of soils and the distribution of vegetation. The fourth, land cover, is the vegetation that is found in a location as a response to the physical factors.

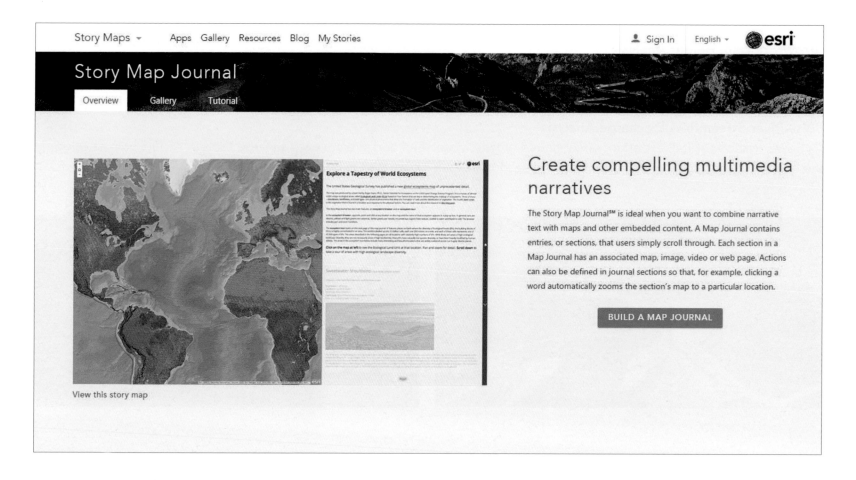

View this story map

# Coastal Community Vulnerability Assessment in Talbot County, Maryland

National Oceanic and Atmospheric Administration, National Centers for Coastal Ocean Science

This map and associated data products provide decision-makers with improved information about existing vulnerabilities and climate risks to support preparedness, response, recovery, and resiliency. Coastal communities face impacts from climate-driven flood hazards and often need assistance in prioritizing locations for management. An integrated vulnerability assessment of Talbot County, Maryland, identified priority adaptation locations. These maps provide a deeper understanding of the range of vulnerabilities and climate risks facing one coastal community; they can support siting adaptation projects to reduce the impacts of climate change.

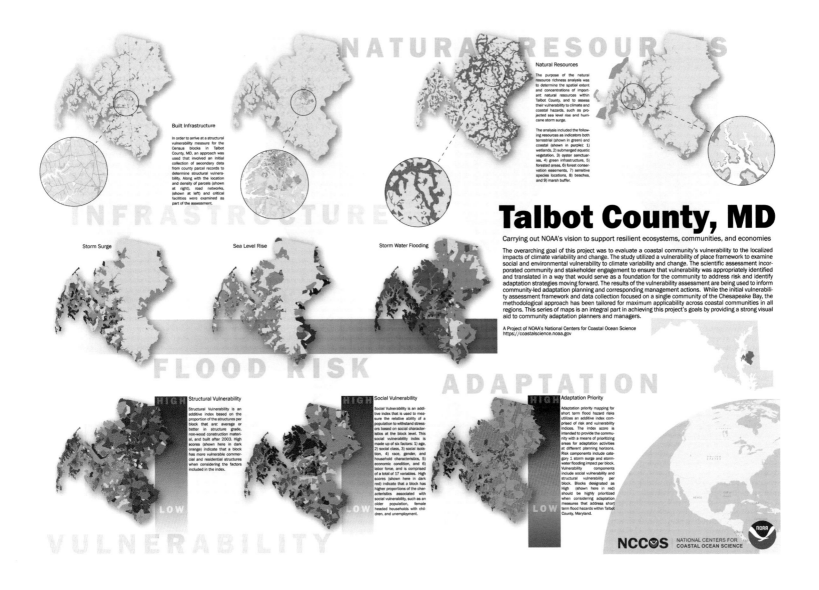

# Smart Location Calculator

US General Services Administration

The General Services Administration (GSA) and Environmental Protection Agency (EPA) partnered with Esri's design team to develop a custom user interface that used an EPA data model to calculate a location sustainability score for any address in the country. The tool is now public and available to anyone interested in choosing sustainable locations for a workplace. They can compare sites and incorporate sustainable site selection data into decisions. The federal government had sustainable location policies/mandates to consider the impacts of site selection on employee and visitor greenhouse gas emissions, but relevant public data was limited. The Smart Location Calculator allows users to compare the location efficiency of sites and easily identify workplace locations that are most likely to limit commuting- and travel-related emissions.

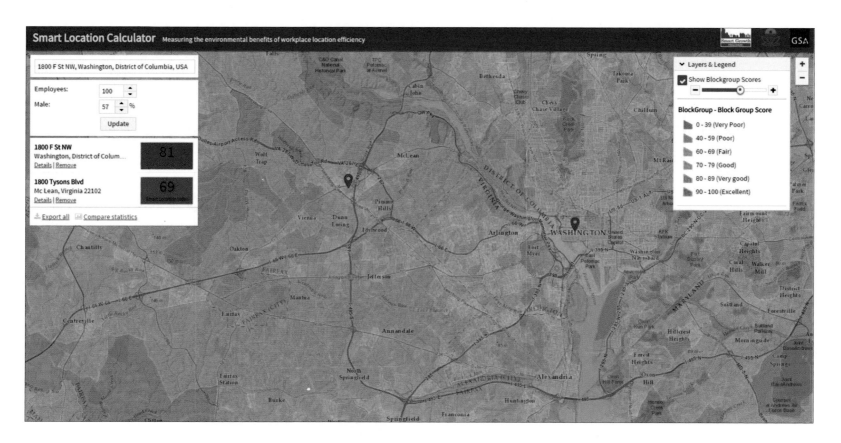

# Monarch Butterfly Conservation

US Fish and Wildlife Service

The US Fish and Wildlife Service has the responsibility to conserve, protect, and enhance fish, wildlife, and plants in their habitats for the continuing benefit of the American people. The environmental legacy passed on to future generations largely depends on protecting and restoring habitat that plants and animals depend on for their survival. The monarch butterfly is one of the most recognizable species in North America. It undertakes one of the world's most remarkable and fascinating migrations, traveling thousands of miles over many generations from Mexico, across the United States, to Canada. Many conservation, education, and research partners from across the United States are working together to conserve the monarch migration.

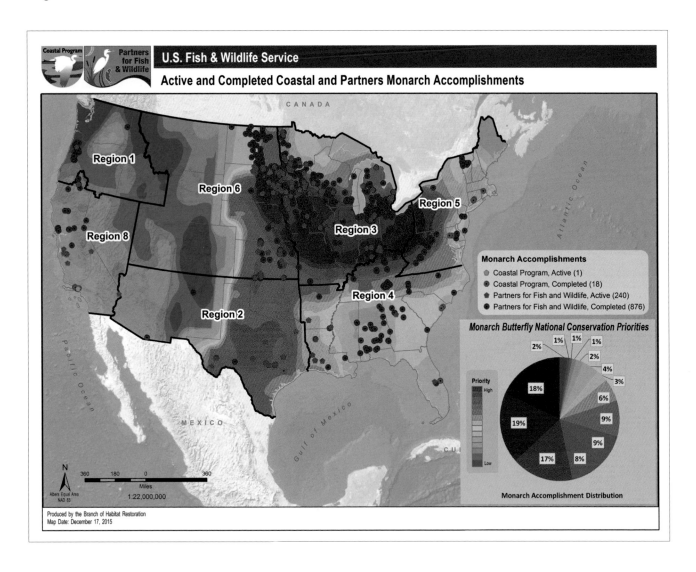

# Steens Mountain Wilderness Area

US Bureau of Land Management, Oregon State Office

The US Congress designated the Steens Mountain Wilderness in 2000, and it now has over 170,200 acres. All this wilderness is located in Oregon and is managed by the Bureau of Land Management (BLM). Steens Mountain is located in Oregon's high desert and is one of the crown jewels of the state's wild lands. It is some of the wildest and most remote land left in Oregon.

Within this area, cooperative and innovative management projects will be maintained and enhanced by the BLM, private landowners, tribes, and other public interests. Sustainable grazing and recreational use, including fishing and hunting, will be continued where consistent with the purpose of the Wilderness Act of 1964. A land exchange provision blocks up nearly 100,000 acres of livestock-free wilderness within the designated 175,000-acre Steens Mountain Wilderness. This land, at the top of the Steens Mountain, is the most sensitive to disturbance and will be managed to safeguard the pristine environment.

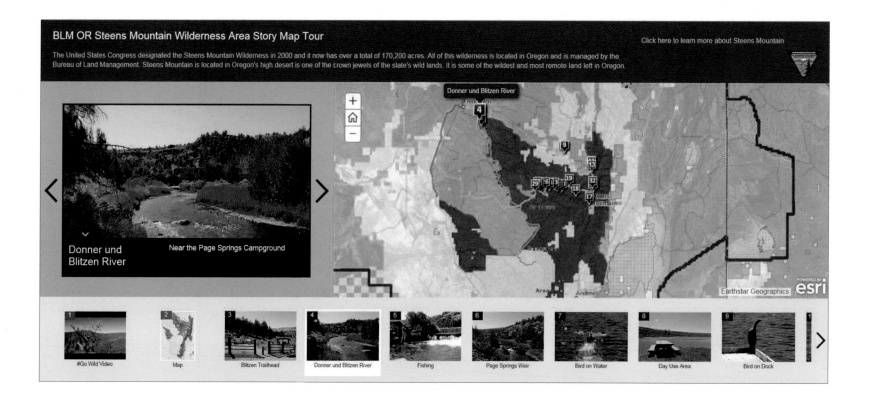

# Lewis and Clark's Scientific Discoveries—Animals

National Park Service

President Thomas Jefferson instructed Meriwether Lewis to explore the Missouri River and find "the most direct and practicable water communication across this continent." In addition, Jefferson asked that Lewis describe "the animals of the country generally, and especially those not known in the US the remains and accounts of any which may be deemed rare or extinct." This story map displays the animal species that explorers Lewis and Clark encountered and first described for science during the Corps of Discovery Expedition of 1804–1806.

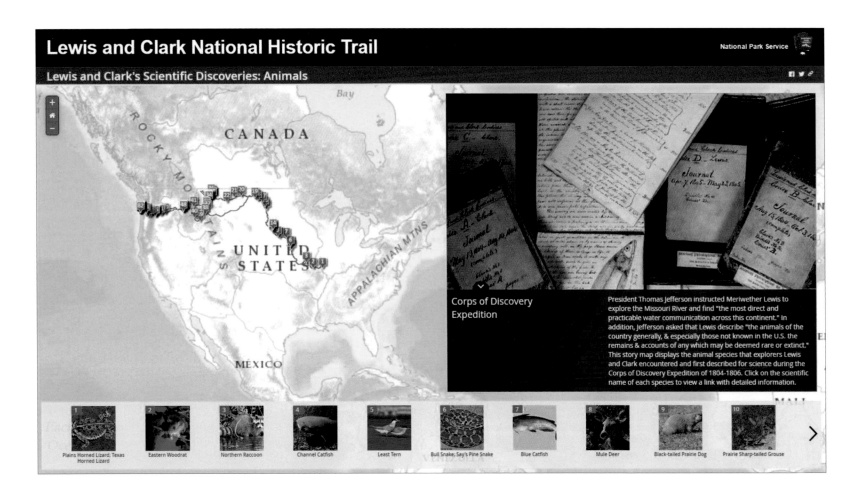

# US Department of Agriculture CarbonScapes

US Department of Agriculture, Natural Resources Conservation Service, National Soil Survey Center Geospatial Research Unit

The US Department of Agriculture's (USDA) CarbonScapes provides a helpful and easy-to-use web map application to educate and answer questions about USDA inventory, modeling, and mapping of terrestrial biosphere carbon in the landscape. Benefits include improved understanding and decision-making for landowners and USDA staff relative to land-use scenarios that enhance carbon sequestration rate potentials for offset of greenhouse gas emissions. The app provides identification and quantification of terrestrial biosphere carbon pools within the landscape using USDA inventoried and modeled sources. Users can answer questions about the mass and stock of various carbon pools within local watersheds or ecoregions and examine USDA data.gov metadata for data and model sources.

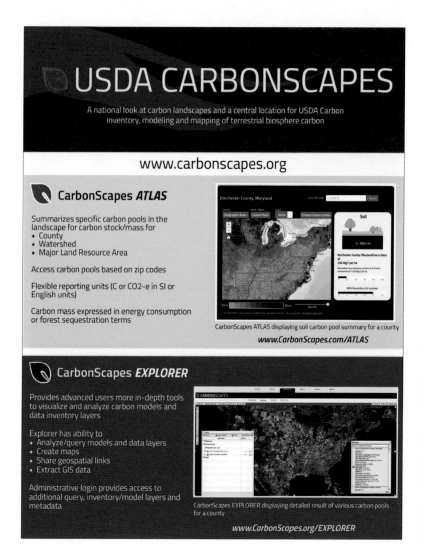

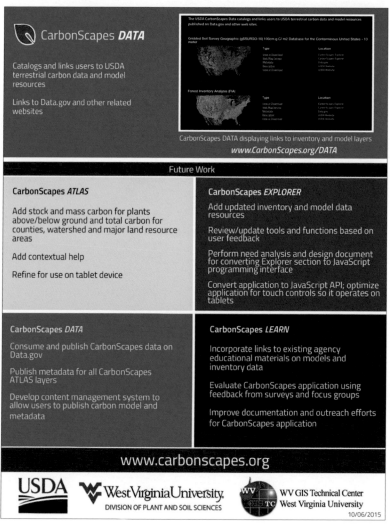

# Methods for Locating Legacy Wells in America's Oldest Oil Field: Oil Creek State Park

US Department of Energy, National Energy Technology Laboratory

This map displays results of an airborne magnetic survey which identified well locations from the 1860s within Oil Creek State Park. The research project was designed to study methods for locating historic wells in Pennsylvania. This is an on-going effort as many of these wells have not been properly plugged and sealed. Open well bores may provide a pathway for subsurface fluids. Remote sensing methods as in this helicopter survey may present a faster and more efficient method for locating historic wells compared to ground surveys.

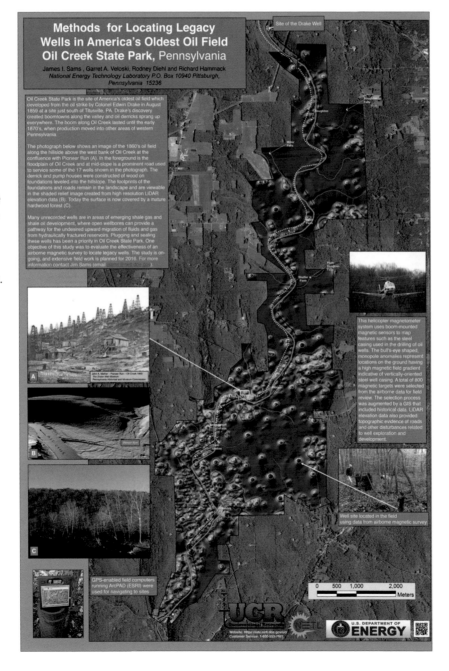

# Celebrating Thirty Years of the Conservation Reserve Program

US Department of Agriculture, Farm Service Agency

This interactive story map provides highlights of the Conservation Reserve Program and is updated in conjunction with feature stories written for the Farm Service Agency's Fencepost Blog. The story map provides improved communication with the public, stimulates support for voluntary conservation practices, and helps ensure long-term program viability. Traditional outreach efforts, including news releases and fact sheets, often fail to engage the general public in conservation education. This interactive map allows the public to see where and how conservation practices are being applied locally through a national program.

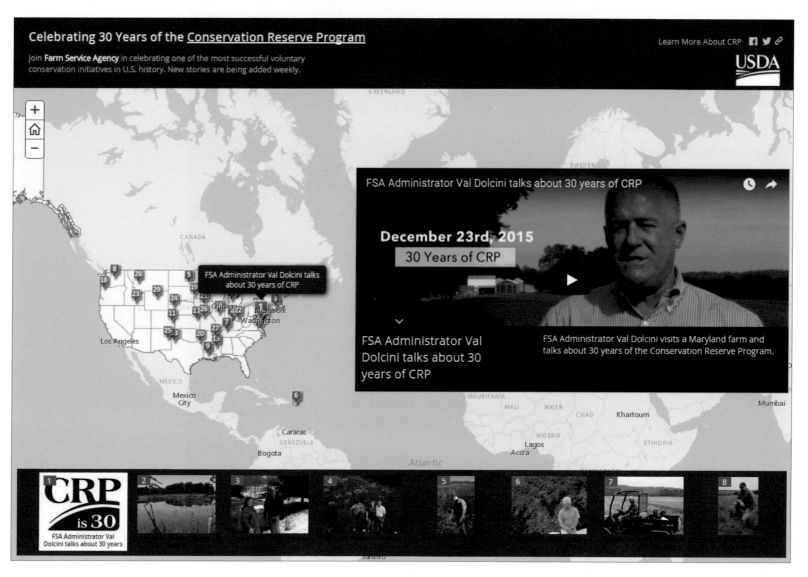

# Visualize Your Water Challenge for High School Students

Nutrient pollution is one of America's most widespread and costly environmental problems, causing algal blooms and hypoxia in local watersheds. To engage high school students in exploring possible solutions to these problems, a coalition of government agencies and private industry created the "Visualize Your Water Challenge."  Students in the thirteen US states bordering he Great Lakes basin and Chesapeake Bay watershed were challenged to use Esri digital mapping technology and open water-quality datasets from the US Environmental Protection Agency (EPA), US Geological Survey (USGS), and others to raise awareness of critical nutrient pollution problems.

Students were charged with creating compelling, innovative, and comprehensive visualizations that inform individuals and communities about nutrient pollution and inspire them to reduce nutrient levels. Submissions were judged on scientific excellence, data analytics, design, and storytelling.  Key collaborators included EPA, USGS, Esri, Chesapeake Bay Foundation, Minnesota Sea Grant, Wisconsin Sea Grant, Center for Great Lakes Literacy, Great Lakes Observing System, and the National Geographic Society, with the assistance of the US Department of Education.

Congratulations to the winners of the Visualize Your Water Challenge and thanks to all of the participants:

**Grand Prize**—Nicholas Oliveira, Washington-Lee High School, Virginia—"Understanding Eutrophication in the Chesapeake Bay"

**Chesapeake Bay Region First Place**—Alex Jin, Poolesville High School, Maryland—"Nutrient Pollution, The Bay's Biggest Threat"

**Great Lakes Region First Place**—Ben Bratton, Father Gabriel Richard High School, Michigan—"Algae Affliction of Lake Erie"

**National Geographic Prize**—Anna Lujan, Washington-Lee High School, Virginia—"Eutrophication in the Chesapeake Bay: Fertilizer and Manure"

**Honorable Mention**—Clara Benadon, Poolesville High School, Maryland—"The Chesapeake Bay:  A National Treasure in Trouble"

**Honorable Mention**—Sam Hull, Poolesville High School, Maryland—"The Bonds of Water"

Visit http://www.esri.com/visualizeyourwater to view all the winning entries.

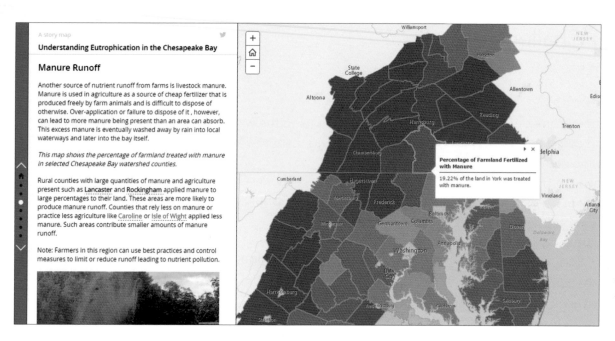

**Understanding
Eutrophication in
the Chesapeake Bay**
Washington-Lee High School
Nicholas Oliveira

**The Bonds of Water**
Poolesville High School/
Sam Hull

## Nutrient Pollution, the Bay's Biggest Threat

Poolesville High School/Alex Jin

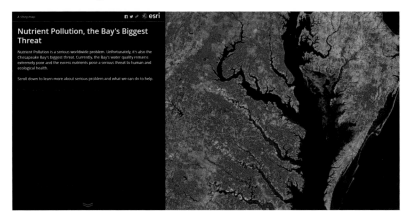

## The Chesapeake Bay - A National Treasure in Trouble

Poolesville High School/Clara Benadon

## Eutrophication in the Chesapeake Bay: Fertilizer and Manure

Washington-Lee High School/ Anna Lujan

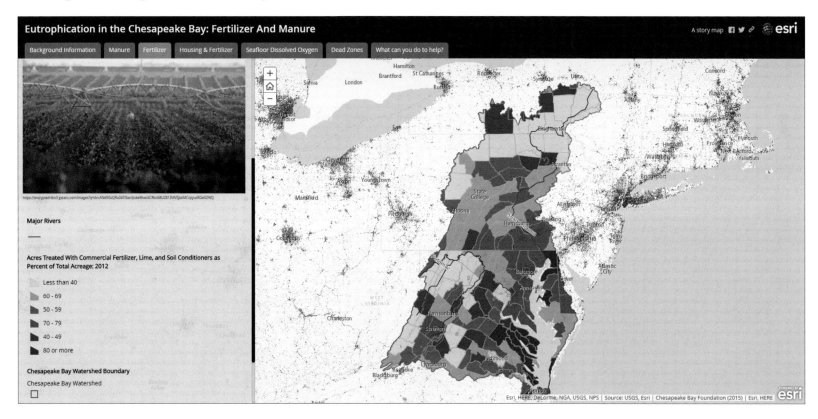

# SOCIOECONOMICS

"This (infrastructure) modernization allows us to support the department's vision of safety, innovation, and opportunity with data, with GIS, and with the cloud. We're focused on how the transportation system can be improved to connect our people with jobs, education, health care, and other important services."

**Dan Morgan**
Chief Data Officer
US Department of Transportation

"Providing relevant information is important across our entire spectrum of users. Embedded maps deliver timely, interactive content to strengthen the media's ability to share our stories."

**Christopher Oswalt**
Research Forester
US Department of Agriculture Forest Service

# SOCIOECONOMICS

Vibrant communities afford social and economic opportunities to all citizens and serve as a basis for a healthy nation. GIS is making it possible for government organizations and officials to better understand the current social and economic health of their communities.

Intelligent maps derived from GIS demonstrate how the world around us and its elements influence the local economy and the social determinants of health. Equipped with geographic insight, decision-makers at all levels of government use GIS to conduct real-time analysis and identify trends that affect supply chains, tourism, public health, education, and other critical components of a functioning society.

From seeing energy consumption in a geographic context, to visualizing access to care among constituents by overlaying demographic data on a map, policy makers rely on GIS to better understand place-based impacts to local, regional, and state economic and social vitality. In turn, legislative leaders are better prepared to make data-driven prescriptive and predictive decisions that bolster the country's social and economic health at large.

# Zika Maps

Centers for Disease Control and Prevention/Food and Drug Administration/National Institutes of Health

Organizations and agencies such as the Centers for Disease Control and Prevention (CDC), Food and Drug Administration (FDA), and the National Institutes of Health (NIH) are working diligently with federal partners and stakeholders in an effort to combat Zika, a virus spread to people primarily through the bite of an infected mosquito. The maps shown here represent a view into the historical timeline, the current state, and moving forward. Also included in this sampling is a story map about analyzing vector-borne disease with the ArcGIS Predictive Analysis Tools. These analysis tools can be used to analyze, create, and disseminate maps that act as decision support for all kinds of vector control activities.

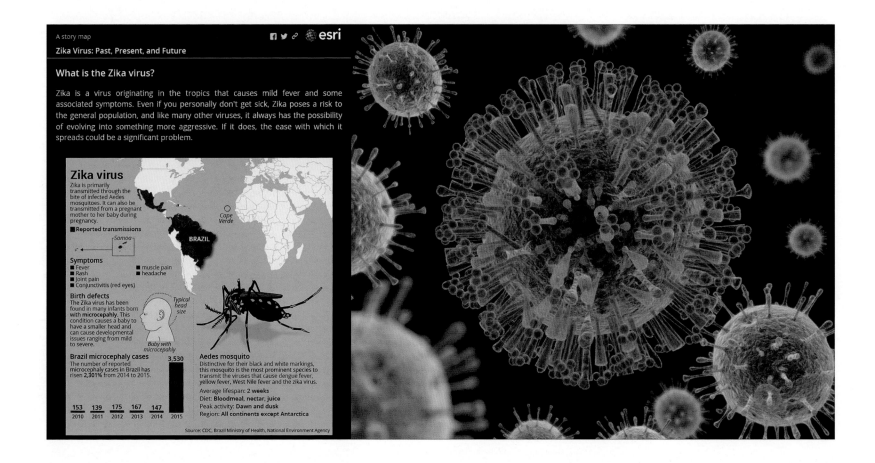

A story map

# Location Platform for Vector-borne Disease Surveillance and Control

Many different organization and agencies are tasked with protecting the health of residents by managing operations to control mosquitoes and other animals or insects that can be vectors transmitting disease. The ability to respond to potential issues and outbreaks quickly and effectively means a world of difference to these agencies and the communities they serve. Outbreaks of diseases such as Zika virus, West Nile and dengue fever have captured the public's attention and highlight growing public concern about vector-borne diseases

Vectors

At Esri, we've found that these organizations typically face 5 major challenges that stand in the way of protecting the health of residents. This Map Journal provides brief proof points to illustrate how the ArcGIS Platform helps overcome these challenges and achieve organizational goals.

## Multiple Systems

ArcGIS provides a single system of record for Call Center staff to collect and verify location information and capture other details from incoming requests. ArcGIS enables both residents and internal staff in responsible organizations to submit requests and report issues from their homes, office or in the field using a variety

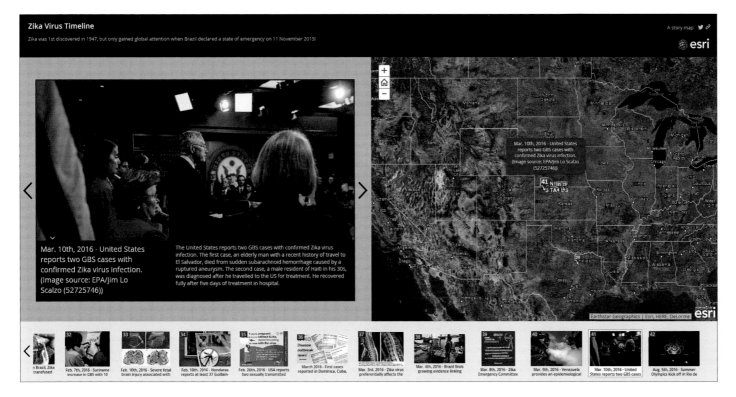

# Zika Virus Timeline

Zika was 1st discovered in 1947, but only gained global attention when Brazil declared a state of emergency on 11 November 2015!

A story map

Mar. 10th, 2016 - United States reports two GBS cases with confirmed Zika virus infection. (Image source: EPA/Jim Lo Scalzo (52725746))

The United States reports two GBS cases with confirmed Zika virus infection. The first case, an elderly man with a recent history of travel to El Salvador, died from sudden subarachnoid hemorrhage caused by a ruptured aneurysm. The second case, a male resident of Haiti in his 30s, was diagnosed after he travelled to the US for treatment. He recovered fully after five days of treatment in hospital.

Mar. 10th, 2016 - United States reports two GBS cases with confirmed Zika virus infection. (Image source: EPA/Jim Lo Scalzo (52725746))

# Integrated Biosurveillance: Alert and Response Operations Products

Armed Forces Health Surveillance Branch

The Armed Forces Health Surveillance Branch (AFHSB) operates under the Defense Health Agency as the central epidemiologic resource and global health surveillance proponent of the US Armed Forces. AFHSB monitors biosurveillance data sources and communicates routinely with the Department of Defense (DoD), US government interagency, and nongovernmental organizations and international partners to detect and report all-hazard events relevant to the health of all DoD personnel, including dependents and beneficiaries.

The US map depicts states that have reported imported Zika cases and the estimated temporal range of mosquito vectors and transmission suitability between the months of July and November. It provides situational awareness for DoD decision-makers about an ongoing vector-borne disease outbreak of military relevance. The other map illustrates Angola's largest yellow fever outbreak in thirty years which spread into the Democratic Republic of the Congo (DRC). Exported cases related to this outbreak have also been reported from Kenya and China. Additionally, Uganda has reported local transmission of the virus that is unrelated to the outbreak in Angola and DRC.

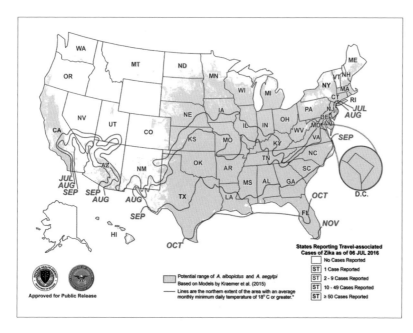

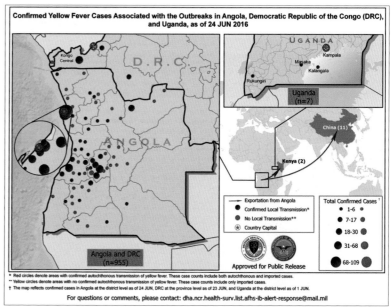

# Veteran Homelessness

Office of US Congressman Mark Takano

California Congressman Mark Takano used this map of veteran homelessness as an example for other offices on the House Committee on Veterans Affairs, of which he was acting ranking member, to use and better display veterans' data for their constituents. The office also shared the map publicly through social media. The green states are where homelessness has decreased and the pink states are where homelessness has increased. This map provides a good snapshot of where progress has been made in decreasing veteran homelessness and where there are places for improvement.

# Our Nation's Veterans

US Census Bureau

This map display shows a variety of veterans-related data visualizations based on themes contained in the US Census Bureau's 2013 American Community Survey (ACS) 5-Year Estimates dataset. It promotes the vast amount of data collected by the ACS that is available, for free, to the public. This display enables people to easily visualize this subset of the 5-Year Estimates data so they can better understand these themes.

# Second Amendment Sentiment

Office of US Senator James E. Risch

To best represent Senator James E. Risch's Idaho constituents, his office reviews their correspondence as a guiding measure on issues that matter most to constituents. The office also boasts a rapid response rate to all forms of correspondence. ArcGIS software enables the office to better understand constituent concerns from all parts of the state. The swipe map displayed shows constituent sentiment on Second Amendment issues in 2016.

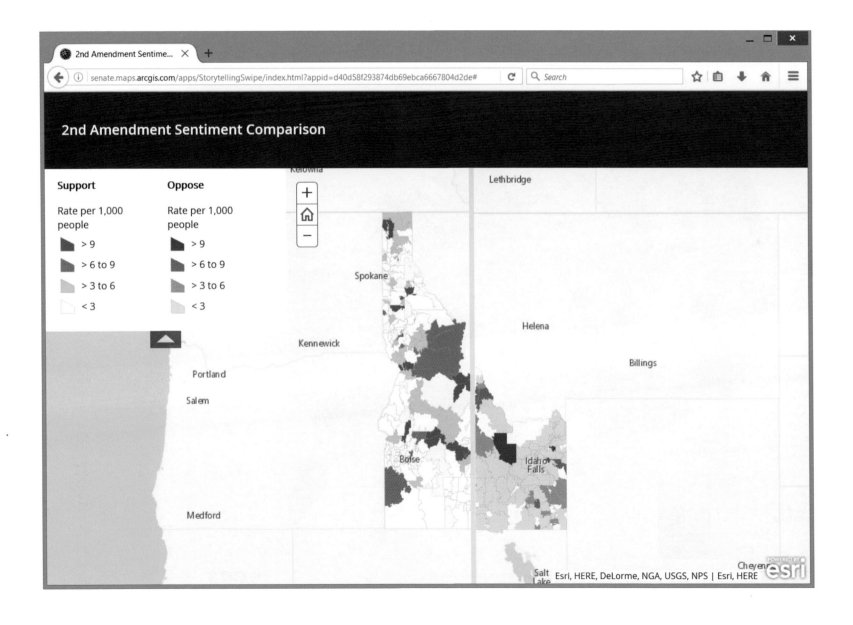

# Percentage of Medicare Beneficiaries Enrolled in Medicare Advantage and TRICARE for Life by County

Congressional Research Service, Library of Congress

The Congressional Research Service (CRS) of the Library of Congress provides authoritative, confidential, objective, and nonpartisan policy and legal analysis to Congress. CRS created a map to illustrate patterns of enrollment for two programs: Medicare Advantage and TRICARE for Life. Under the Medicare Advantage program, eligible Medicare beneficiaries can receive their covered health-care services through a private insurance plan. TRICARE for Life is supplemental health insurance coverage for certain Medicare-eligible military retirees. For both of these programs, enrollment varies geographically. The map shows the combined enrollment of both programs by county.

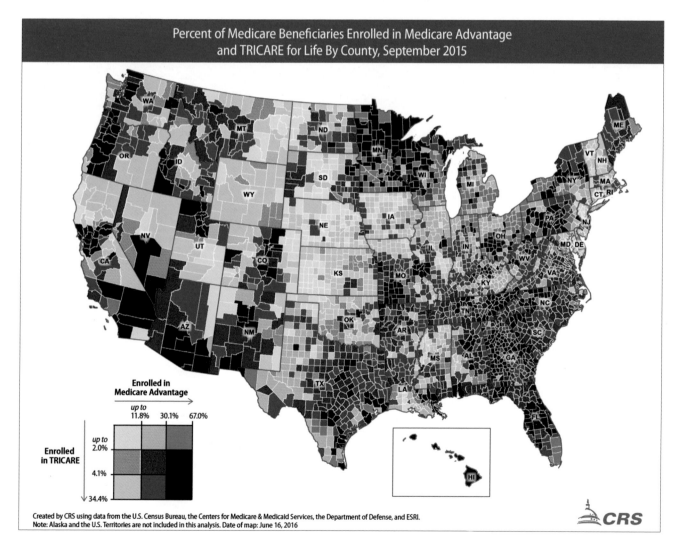

# Multi-Resolution Data Store Automated Map Output

## Ordnance Survey Ireland

This map demonstrates the achievement of 100 percent automation of data and cartographic generalization, plus related critical map element production (indexes, supporting text, etc.) This map uses an existing definitive core data holding (Prime2) and enables the delivery of value by Ordnance Survey Ireland (OSi) from that data holding to all OSi stakeholders. This map represents the key targets for automation and efficiencies within a national mapping organization as it harnesses existing expertise and provides an automation capability through ArcGIS. The technology, workflow, and process behind the production of this map will allow OSi to automatically refresh existing products and generate new products and services such as open data and open web services.

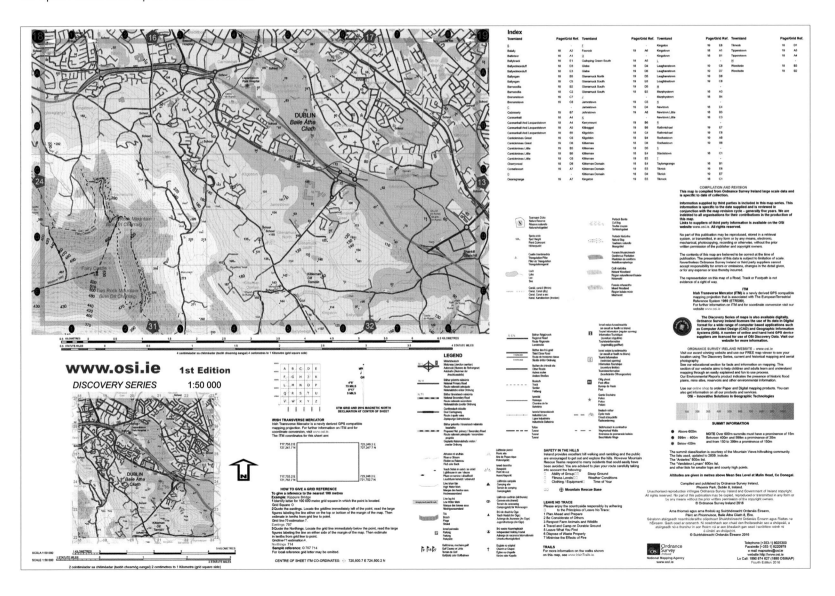

# Visualizing Which Provider to Pursue in a Case of Health-Care Provider Fraud

US Postal Service, Office of Inspector General

After some investigation, five health-care providers were determined to be linked and involved in fraudulently charging the US Postal Service (USPS). Which one has had the biggest impact and should be the focus of prosecution? This map helps the agent make the case that provider 5 should be prosecuted because of the highest impact for the number of claims and the span across the nation. This map aided the agent in making a quick decision so the provider could be stopped as soon as possible and USPS could stop paying the provider for false charges.

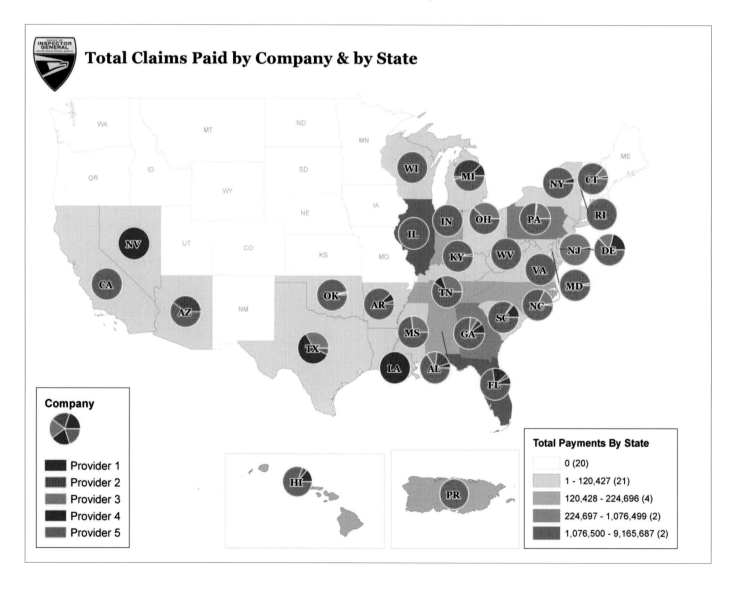

# Visualizing Where Mail is Delayed in Processing Plants

US Postal Service, Office of Inspector General

The US Postal Service (USPS) wanted to know which processing plants have higher rates of delayed mail, where they are located in relation to one another, and if any of these plants show recurring delays. The solution was a visual display that allowed auditors to track the delayed mail over time so they could see the changes and look for plants that continuously have delayed mail and look at the type and volume of mail that is delayed. By visualizing this over time, auditors can begin to look for plants requiring site visits to determine why mail is being delayed—such as understaffing for the volume of mail processed—and then work with USPS to find a solution.

# Social Security's Old-Age, Survivors, and Disability Insurance Beneficiaries and Benefits by State (2014)

Social Security Administration, Office of the Chief Strategic Officer

These maps provide Social Security information on people receiving Old-Age, Survivors, and Disability Insurance benefits, by state in 2014. The goal of the Social Security Administration is to provide the public with a deeper understanding of its programs through geographic representations. Geospatial visualizations help the public understand information about the data. These maps visualize complex data and make it easier for the public and Capitol Hill staff to understand information about Social Security programs, increasing the transparency of the agency's data.

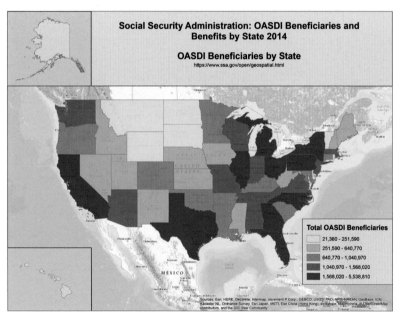

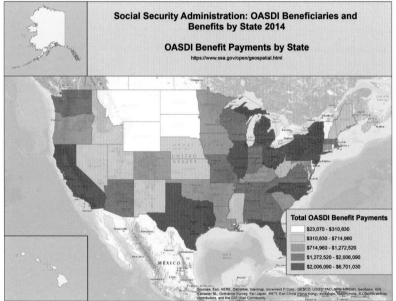

# Counties with No Opioid Treatment Program Facilities, by Level of Urbanization (2014)

White House Office of National Drug Control Policy

This map shows the level of urbanization, based on the Urban-Rural Classification Scheme for Counties by the National Center for Health Statistics, of counties in the United States that lack opioid treatment facilities. Decisions can be made to try to promote the availability of opioid treatment facilities in areas that lack them, particularly in rural areas. This map indicates not only the locations of counties without opioid treatment facilities, but also the level of urbanization of those counties, i.e. rural versus urban areas.

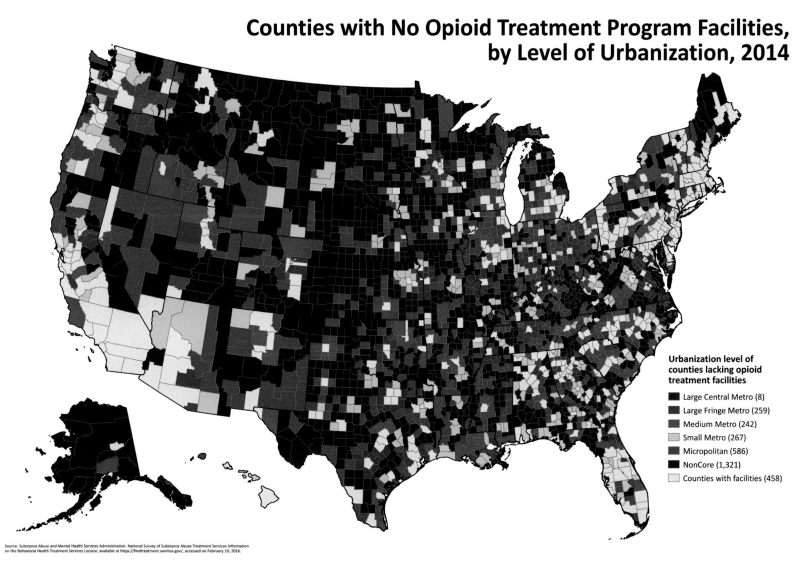

**Counties with No Opioid Treatment Program Facilities, by Level of Urbanization, 2014**

**Urbanization level of counties lacking opioid treatment facilities**

- Large Central Metro (8)
- Large Fringe Metro (259)
- Medium Metro (242)
- Small Metro (267)
- Micropolitan (586)
- NonCore (1,321)
- Counties with facilities (458)

Source: Substance Abuse and Mental Health Services Administration. National Survey of Substance Abuse Treatment Services information on the Behavioral Health Treatment Services Locator, available at https://findtreatment.samhsa.gov/, accessed on February 10, 2016.

# Drug Overdose Deaths Involving Heroin and Other Opioids, by County (2010-2014)

White House Office of National Drug Control Policy

This map uses US Centers for Disease Control and Prevention mortality data to show the rate of drug overdose deaths involving heroin and other opioids for the five-year period of 2010–2014. Maps are used for bringing problems like this to the attention of policy makers and the public. The opioid epidemic is being quantified in this map, showing areas where people are dying as a result of illicit drug use. This map shows areas where the consequences of illicit opioid use are being felt most heavily, providing a good tool to determine where to concentrate resources for treatment and prevention.

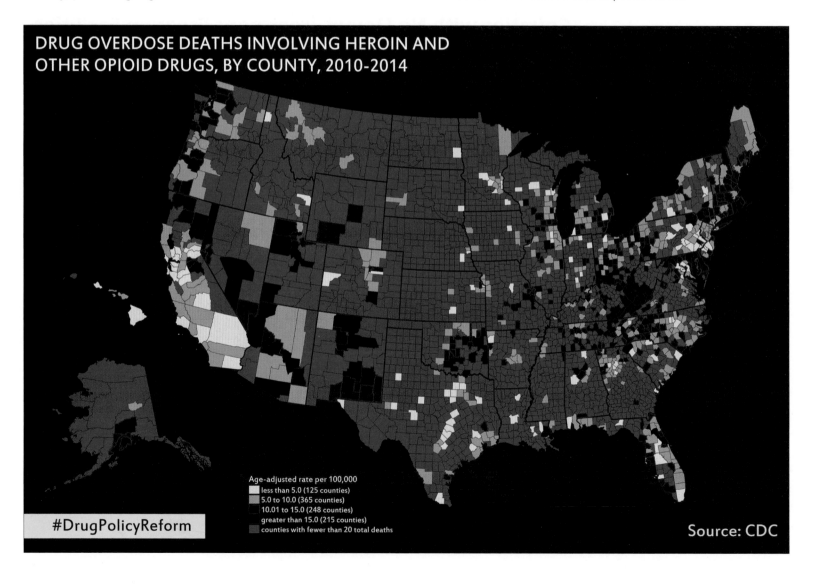

# Laws Increasing Access to Naloxone, a Life-Saving Overdose Reversal Drug

White House Office of National Drug Control Policy

This map shows how state-level legislation has been changed to make the overdose-reversal drug Naloxone more available. The map shows one potential solution to the consequences of the opioid epidemic being addressed through the availability of an important life-saving tool. Overdoses can be reduced by allowing more laypersons to administer Naloxone to overdose victims.

27 STATES AND D.C. HAVE CHANGED THEIR LAWS TO INCREASE ACCESS TO NALOXONE, A LIFE-SAVING OVERDOSE REVERSAL DRUG.

#DrugPolicyReform

SOURCE: THE NETWORK FOR PUBLIC HEALTH LAW

# Medicaid Expenditures on Sovaldi and Patients Treated (2014)

Office of US Senator Ron Wyden

US Senators Ron Wyden (D-Oregon) and Chuck Grassley (R-Iowa) collected Medicaid and treatment data as part of a bipartisan investigation into the pricing strategy of Sovaldi and its impact on the US health-care system. Sovaldi was developed by Gilead Sciences, Inc. to treat the hepatitis C virus (HCV). This map illustrates the data collected showing that in 2014, Medicaid programs spent $1 billion on Gilead's HCV drugs, yet more than 97 percent of estimated Medicaid patients with HCV went untreated.

# Bullseye Mapping Tool

US General Services Administration

The Bullseye Viewer is a mapping tool to determine real estate markets, submarkets, flood zones, seismic zones, and central business districts. It improves efficiency for lease evaluations and supports the goal of the General Services Administration to negotiate leases at or below market rate. The tool allows users to easily identify multiple applicable markets and zones by drawing the delineated area just once.

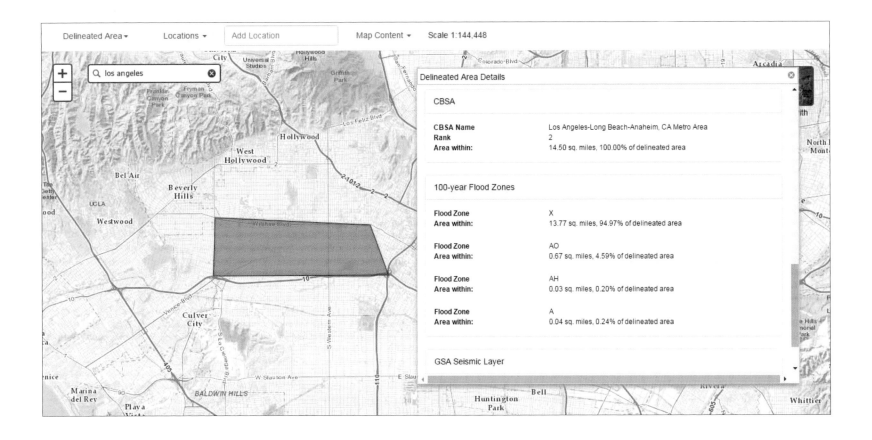

# General Services Administration Fiscal Year 2017 Budget Request

US General Services Administration

This story map shows General Services Administration (GSA) projects that design, construct, repair, and manage its real estate portfolio in communities around the country. These maps support the president's Open Government Initiative, displaying GSA's future projects in a dynamic way and allowing the public to understand GSA's real estate investments in their communities.

GSA wanted to share the president's budget request for its 2017 fiscal year capital investment program so the American people could understand the potential impacts of this significant federal investment. GSA built a story map of projects included in the president's budget. The maps provide a national view as well as details from border station, campus, courthouse, and office building projects.

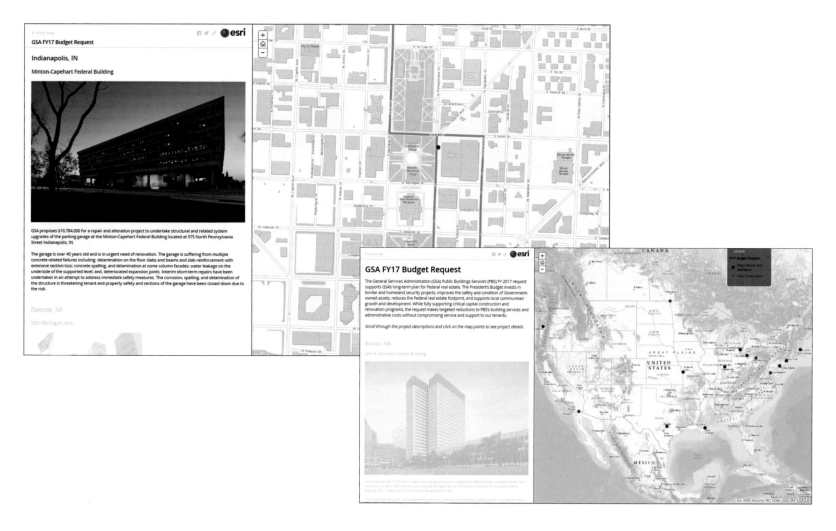

# Voter Turnout in Chicago (2008-2015)

University of Chicago

Each map displays hot and cold spots of voter turnout in Chicago by voting precinct over six general elections. They provide an explanation of voting trends and analysis of voter behavior according to geography and demographics, displaying where registered voters show up at the polls on a regular basis. Each map correlates voter turnout with other demographic variables to show what the motivating factors are in a general election in Chicago.

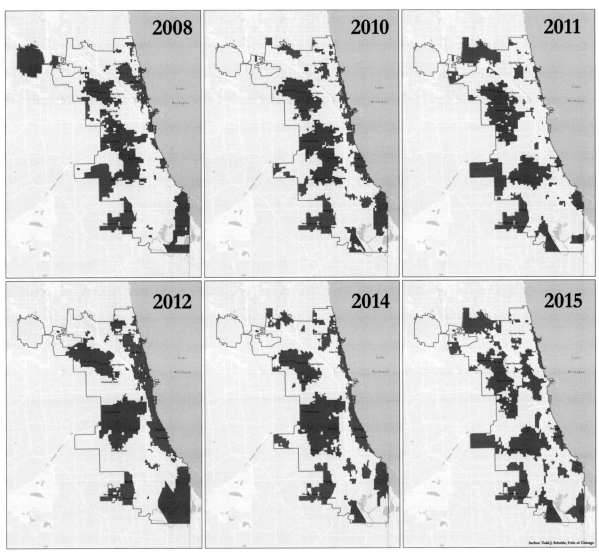

## Voter Turnout in Chicago (2008-2015)

General elections in Chicago have various voter turnouts depending on what offices are up for re-election and who the candidates are. Registered voters in Chicago come out in greater numbers in elections that coincide with higher offices like President of the United States or Governor of Illinois. Municipal general elections can still draw respectable numbers of voters to the polls depending on the candidates and the office. However, there is an underlying issue that often goes overlooked when analyzing the political landscape in Chicago. Who is going to vote (or not vote) in a general election in Chicago regardless of the issues or candidates?

There are parts of the city of Chicago that consistently have high and low voter turnout. By analyzing six consecutive general elections, a definitive pattern emerges as to where any candidate running for any office has a high or low likelihood of securing a large number of votes.

The data in this report comes from the Chicago Board of Elections (http://www.chicagoelections.com), which tracks and maintains voting records for all elections within the City of Chicago. Using precinct level data on the number of registered voters and ballots casts from six general elections, we conducted a Hot Spot analysis of voter turnout by precinct. This spatial analysis method, also known as Getis-Ord Gi*, was conducted using geographic information systems (GIS) software, ESRI's ArcGIS Desktop 10.3. The mapping software isolated areas that had either relative high or relative low voter turnout rates when compared with the rest of the city. The precincts in red highlight clusters of high voter turnout, "a hot spot", while precincts in blue highlight clusters of low voter turnout, "a cold spot".

From year to year, there are undeniable patterns in the spatial data. High turnout occurs in every election in neighborhoods like Mt. Greenwood, Beverly, Morgan Park, Kenwood, Hyde Park, Lincoln Square, Ravenswood, eastern parts of Lakeview, and northern portions of Streeterville. While there is consistent low voter turnout in Belmont Cragin, Hermosa, South Lawndale, the northern portion of Humboldt Park, eastern portion of Logan Square, along with areas in Chicago Lawn and New City neighborhoods.

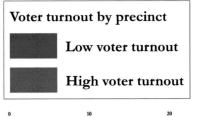

**Voter turnout by precinct**

Low voter turnout

High voter turnout

# Averaged Eligibility Map

US Department of Agriculture, Food and Nutrition Service

The Averaged Eligibility Map shows if a summer meal site or day care home is located in a potentially eligible location based on the USDA Food and Nutrition Service's averaging policy.

Day care homes (participating in the Child and Adult Care Food Program) and summer meal sites (participating in the Summer Food Service Program or the Seamless Summer Option) are considered area eligible if located in a census block group (CBG) that, when combined with one or two adjacent CBGs, has a weighted average of 50 percent or more children eligible for free or reduced-price meals. Each CBG included in the calculation must also have at least 40 percent of children eligible for free or reduced-price meals (FRP). Census tracts may not be combined.

The purpose of this map is to quickly show if a particular summer meal site or day care home is located in a CBG that meets the USDA's policy requirements. A users' guide helps communities interpret the map and use the data provided in the map to ask their state agency to formally designate their site as area eligible to receive funding.

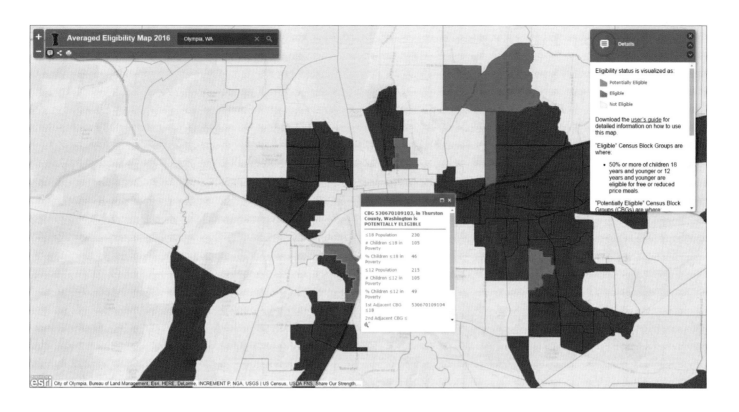

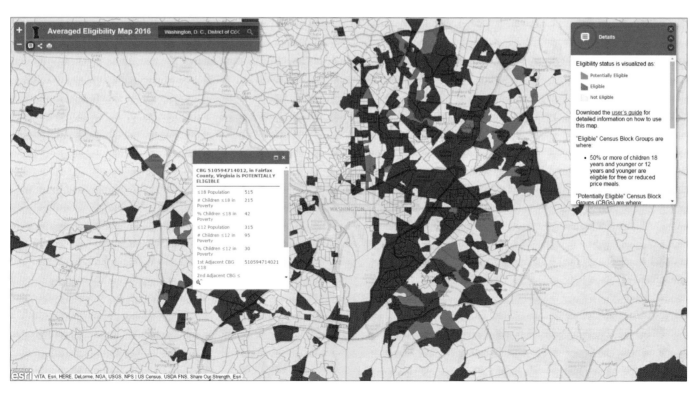

# Highly Pathogenic Avian Influenza Scenarios Affecting Poultry

US Department of Agriculture

Highly pathogenic avian influenza (HPAI) was identified in the United States in December 2014. The US map shows the poultry population inventories and numbers of confirmed HPAI poultry numbers. In April and May, 184 of the 211 commercial cases occurred in the upper Midwest. Putting the confirmed HPAI poultry numbers layer on top of the young chicken numbers layer shows where poultry populations are concentrated and which areas are affected. This information can provide emergency planners with the ability to target geographic areas to reduce the spread of HPAI.

The other map shows the infected, buffer, and surveillance zones established in Iowa as of July 16, 2015. The Animal and Plant Health Inspection Service provided map support as a part of the largest animal disease outbreak in US history. These maps were used for daily situation reports and to communicate with trade partners.

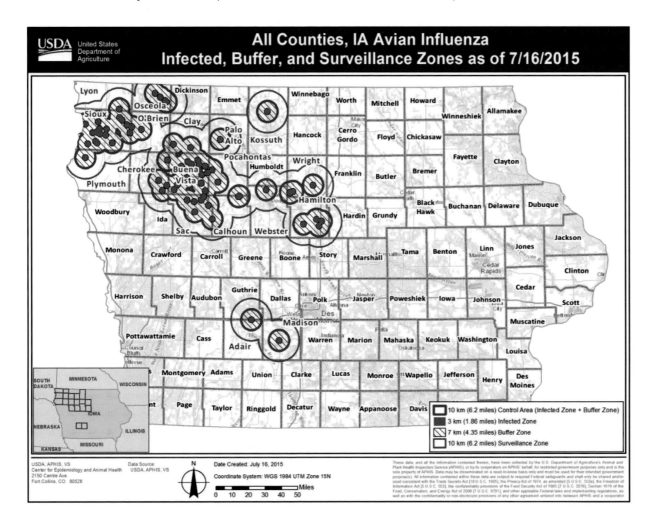

# Crop Migration and Change for Corn, Soybeans, and Spring Wheat in the North Central United States (2006–2015)

US Department of Agriculture, National Agricultural Statistics Service

These maps show mean crop centroids, first-year crop grown, and latest-year crop grown. Identifying where crop planting changes are occurring may help better inform best agricultural practices and improve crop estimation techniques. The maps identify patterns of crop migration and change for corn, soybeans, and spring wheat in the north central states of Minnesota, North Dakota, and South Dakota. The analyses and maps indicate that corn and soybeans have a northwest migration pattern and spring wheat has a west migration trend.

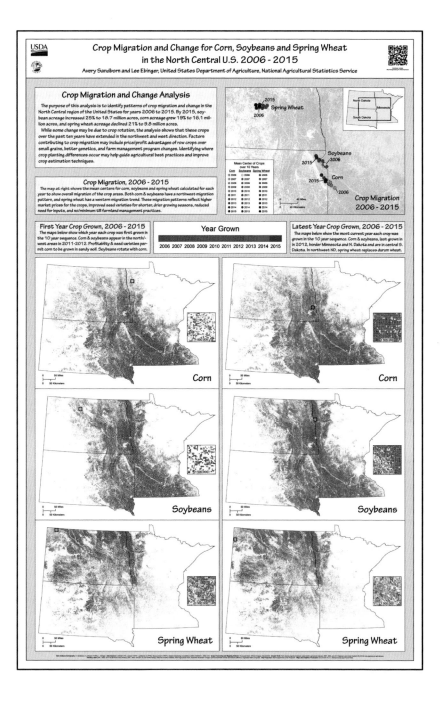

# Using GIS for the Albanian Census

Albanian Institute of Statistics

The 2011 Albania population and housing census was the first large statistical operation conducted in the country with an extensive use of geospatial tools in all the phases of the census exercise: pre-enumeration, enumeration, and post-enumeration phases.

ArcGIS technology allowed the Albanian Institute of Statistics (INSTAT) to create repeatable models to define enumeration areas, as well as create map books for use by field management to guide the process. INSTAT also developed an application based on ArcGIS to monitor in real time the census coverage during data collection operations. The number of interviews completed and the number of households, persons, and housing units enumerated were transmitted from the enumerators and supervisors to the GIS server located at the INSTAT headquarters and mapped with a web feature service. In the post-enumeration phase, INSTAT deployed a web application for data dissemination based on ArcGIS for Server and ArcGIS Online.

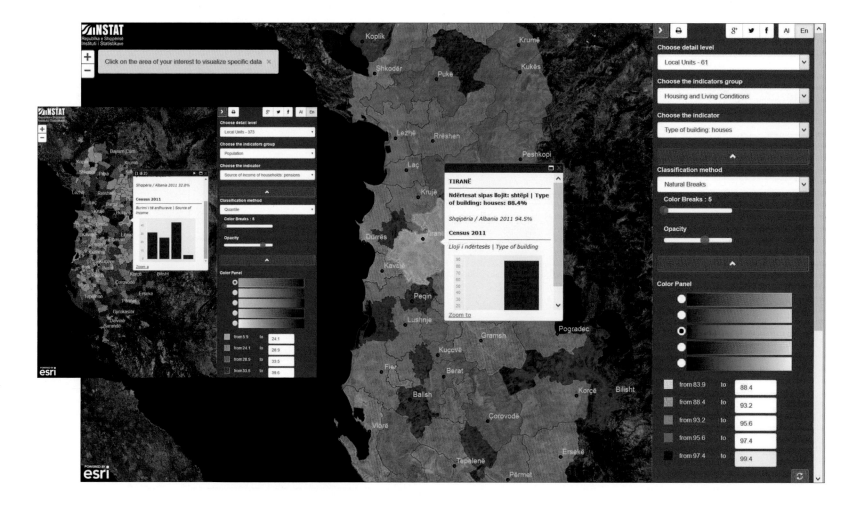

# Geocube: A Flexible Energy and Natural Systems Web Mapping Tool

Mid-Atlantic Technology, Research, and Innovation Center/US Department of Energy, National Energy Technology Laboratory

The National Energy Technology Laboratory's' Geocube was developed to provide a highly customizable, flexible web mapping application with key spatial data, information, tools, and functionality for energy-related activities. Geocube provides a centralized location to perform research, planning, response, assessments, and more without the need for desktop applications.

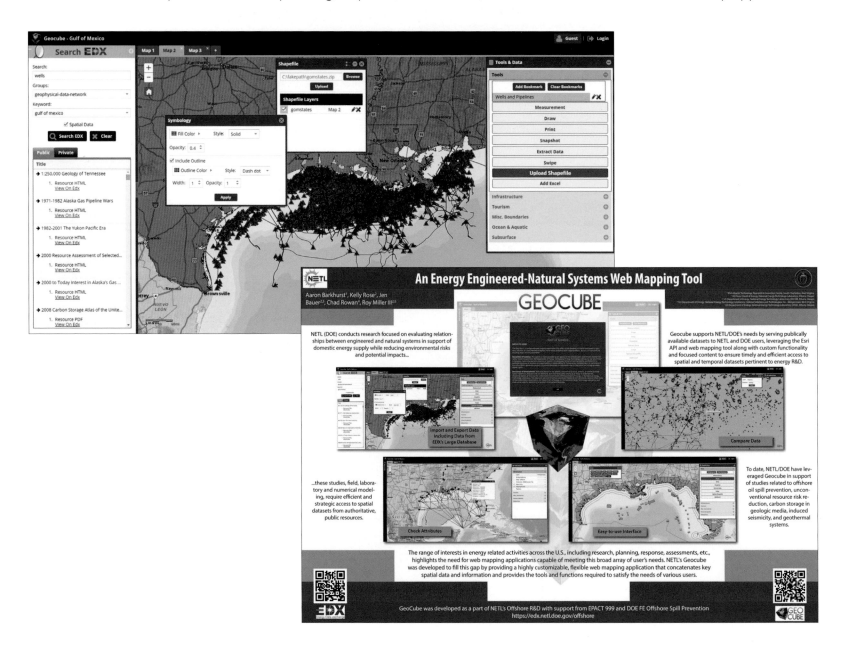

# STATcompiler: The DHS Program

Blue Raster/US Agency for International Development

The Demographic and Health Surveys (DHS) Program of the US Agency for International Development needed a way to present the extensive amount of spatial data that allows the user to view it in multiple formats: in the map, in interactive charts and graphs, and comparing over time. STATcompiler is a focused, interactive application that allows users to see data in multiple formats and download the spatial data directly from the application. Shown in this map is the percentage of women who are literate on a subnational level for the most recent survey year available. This is just one of over three hundred indicators published by the DHS Program. The newest release of STATcompiler is the most modern in design and is completely functional on tablet devices.

# Water More Precious Than Gold: The Story of the Boise Project

US Bureau of Reclamation, Pacific Northwest Region

This story map chronicles the role of the US Bureau of Reclamation in taming the Boise River in Idaho to ensure a reliable supply of irrigation water to Boise Valley farms. It is a tool for public outreach, education, and transparency and uses cutting-edge technology to provide historical information to the public about the crucial role the Bureau of Reclamation played in the settlement of the western United States. A story map is an interactive and entertaining presentation that provides far more information than could be presented in a brochure.

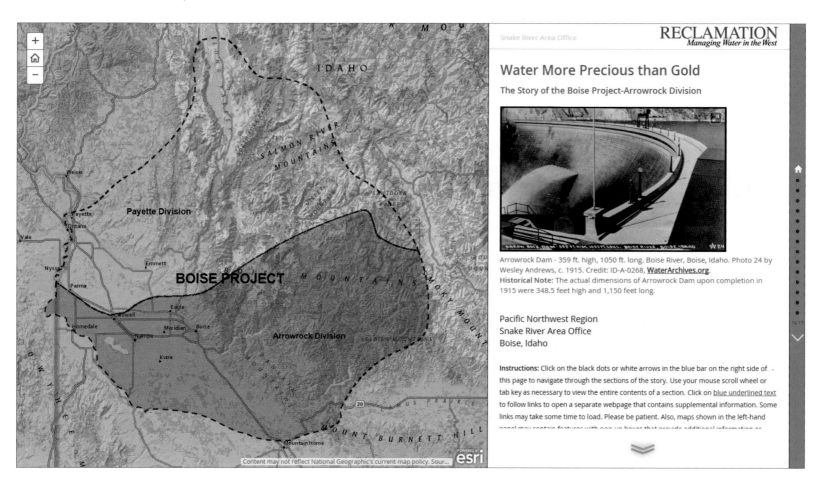

# MapEd

Blue Raster/National Center for Educational Statistics

Displayed are public school points symbolized on National Center for Educational Statistics (NCES) locale codes, overlaid with high school attendance boundaries. These are just two of the many spatial data layers relating to education. Lacking was a repository of spatial education data collected from multiple sources in a viewer that was interactive, all-encompassing, and had the option to view more than one dataset at a time. Blue Raster created an application that is accessible and easy to use so that anyone—from a student to a policy maker—has access to data collected by NCES and the US Census Bureau in one location.

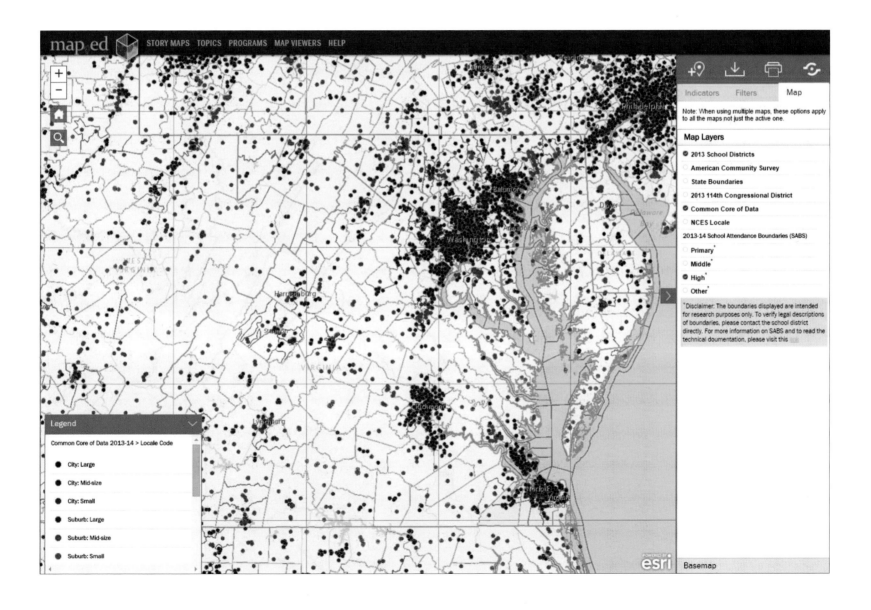

# The Hidden Cost of Suspension

Blue Raster/US Department of Education

This story map shows the percentage of Out-of-School Suspension (OSS) rates for African American students with disabilities by school district, one of four maps created to address OSS across the United States. Getting this data into the hands of policy makers allows for the creation of effective programs. Schools use suspension as punishment, but removing students from the classroom leads to other problems. Such factors as race, gender, and disabilities are often overlooked or not understood. Users can compare maps to find geographic patterns for suspension rates so that programs can be implemented and improved to ensure safe, comfortable learning environments for children.

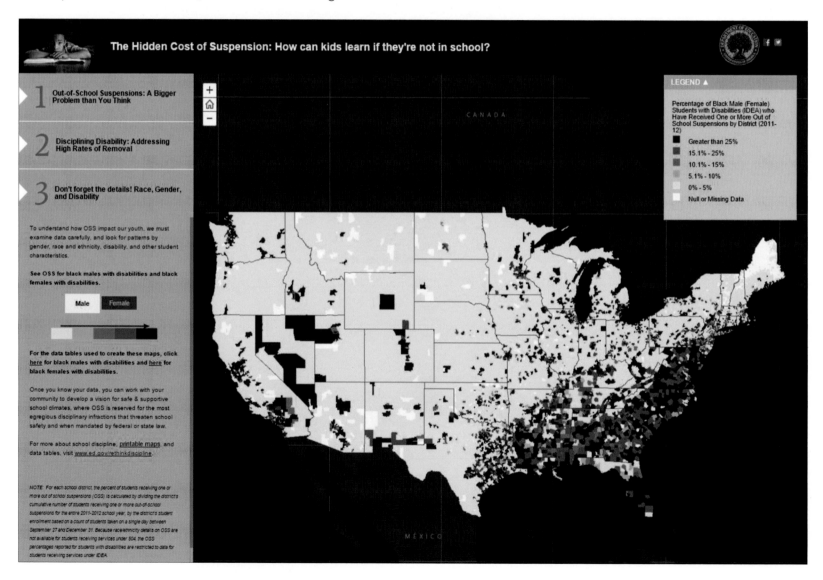

# Naval Station Norfolk Measurements of Energy Reduction

GISInc.

The US Navy is increasing energy awareness with the Navy Shore Geospatial Energy Module (NSGEM), a tool designed to help improve energy efficiency at Navy installations. Energy consumption data for Navy installations comes from a variety of sources and business systems. The NSGEM provides a common interface that allows users to explore the datasets geospatially. The NSGEM assists the Navy with data accuracy, centralized information, transparency, and visibility of information in energy consumption throughout every Navy installation by translating information into visual cues for better data interpretation. This is a view of six different graphs that help the Navy measure energy reduction. The maps provide information to make decisions regarding overconsumption of energy at installations.

# Top Twenty-Five US–International Trade Freight Gateways by Value of Shipments (2014)

US Department of Transportation

This map shows the top twenty-five transportation facilities that move international trade into and out of the United States. The twenty-five gateways are ranked according to the value of shipments in 2014. The map supported a report by the US Department of Transportation's Bureau of Transportation Statistics detailing freight facts.

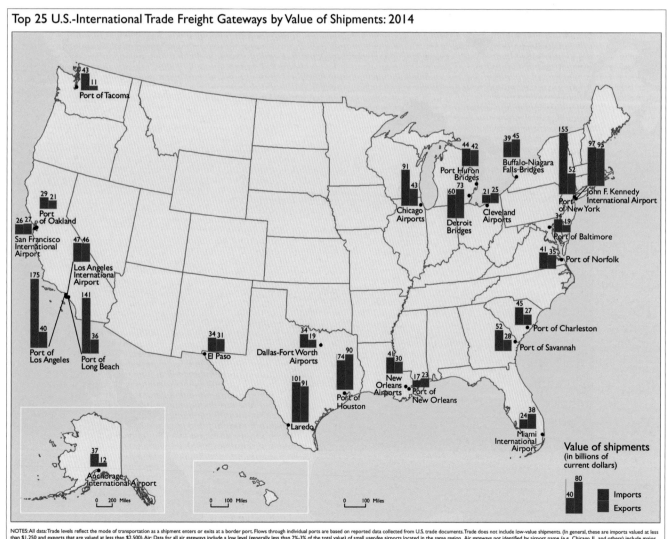

Top 25 U.S.-International Trade Freight Gateways by Value of Shipments: 2014

NOTES: All data: Trade levels reflect the mode of transportation as a shipment enters or exits at a border port. Flows through individual ports are based on reported data collected from U.S. trade documents. Trade does not include low-value shipments. (In general, these are imports valued at less than $1,250 and exports that are valued at less than $2,500). Air: Data for all air gateways include a low level (generally less than 2%-3% of the total value) of small user-fee airports located in the same region. Air gateways not identified by airport name (e.g., Chicago, IL, and others) include major airport(s) in that geographic area in addition to small regional airports. In addition, due to U.S. Census Bureau confidentiality regulations, data for courier operations are included in the airport totals for JFK International Airport, Cleveland, New Orleans, Los Angeles, Chicago, Miami, and Anchorage. To further protect data for individual couriers, data for Memphis is included with New Orleans and data for Louisville is included with Cleveland.
SOURCES: Air: U.S. Department of Commerce, U.S. Census Bureau, Foreign Trade Division, USA Trade Online, October 2015; Water: U.S. Army Corps of Engineers, Navigation Data Center, special tabulation, October 2015; Land: U.S. Department of Transportation, Bureau of Transportation Statistics, North American TransBorder Freight Data, available at www.bts.gov/programs/international/transborder/ as of October 2015.

# The Bartholdi Fountain

House Historian's Office

This map provides a visual tour of the locations proposed for the Bartholdi Fountain when it was moved from the Capitol Lawn in the early 1900s. The House Historian's Office was looking for a convenient way to visually display the many places that were being mentioned in its website's weekly blog. Location reference matched with era-appropriate photography give a sense of each place mentioned in the blog. This map added value to the House Historian's Office blog and the viewers who read it on a weekly basis.

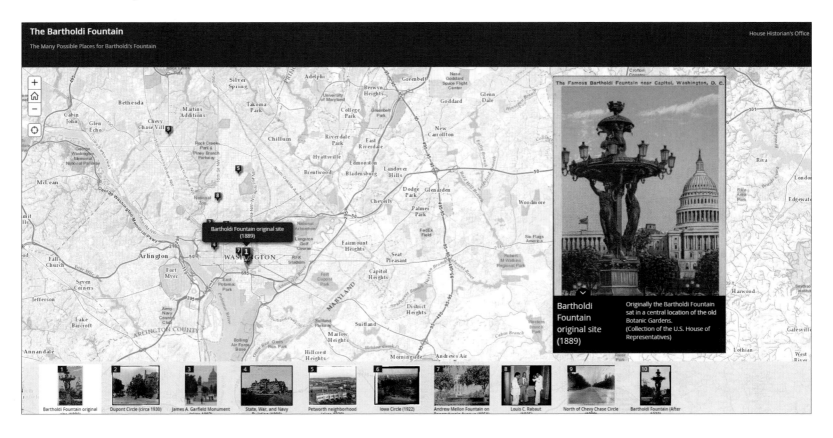

# NATIONAL
# SECURITY

"Providing authoritative, trusted, and open data is fundamental to the DHS mission… We have an opportunity to be open and secure, to empower citizens and communities, to support local law enforcement and first responders, businesses, and the private sector. Homeland security requires a whole-of-nation approach."

**David Alexander**
Geospatial Information Officer
US Department of Homeland Security

"When we match government's purpose, resources, and years of tradecraft with the energy and innovation of industry, good things happen."

**Susan Gordon**
Deputy Director
National Geospatial-Intelligence Agency

# NATIONAL SECURITY

The work of a single agency isn't done in a vacuum. To accomplish the mission of keeping people and property safe, organizations throughout all levels of government must work together to plan for emergencies, manage domestic and international risks, and coordinate response and recovery efforts.

GIS makes it possible for officials in public safety to quickly visualize and assess potential and real threats. By sharing local and state data and visualizing it through a common operating picture on a map, federal agencies are better equipped to develop funding and response strategies that deliver the right resources to state and local organizations.

And it's not just people who work for public safety agencies who are leveraging the power of location to make a difference. During the historic 2016 floods in Baton Rouge, Louisiana, the city's Department of Information Services used GIS and streams of data from 911 calls, search-and-rescue efforts, and other sources to create a regional map showing the impact of the floods.

By pooling data from various government agencies, the department provided a common operating picture for all parties involved in the response—from the Federal Emergency Management Agency (FEMA) to the city's local police force. In addition to supplying public safety personnel with the information they needed to speed response and recovery efforts, the map also served as a public information tool that displayed 311 requests, FEMA flood zones, road closures, and other critical data.

# Louisiana Flood Maps

Federal Emergency Management Agency

A slow-moving weather system brought over 20 inches of rain to parts of southcentral and southeast Louisiana from August 11 to August 15, 2016, causing widespread flooding in the Baton Rouge and Lafayette areas. The flooding had a significant impact on the state as eleven people died, over 40,000 homes were damaged, and over 60,000 people sought federal aid.

**Esri Story Map Journal**

To convey the magnitude of this event and share information on current apps and data, an Esri® Story Map Journal℠ was put together, compiling a lot of information quickly in an easy-to-consume way. This app helped brief decision-makers and elected officials.

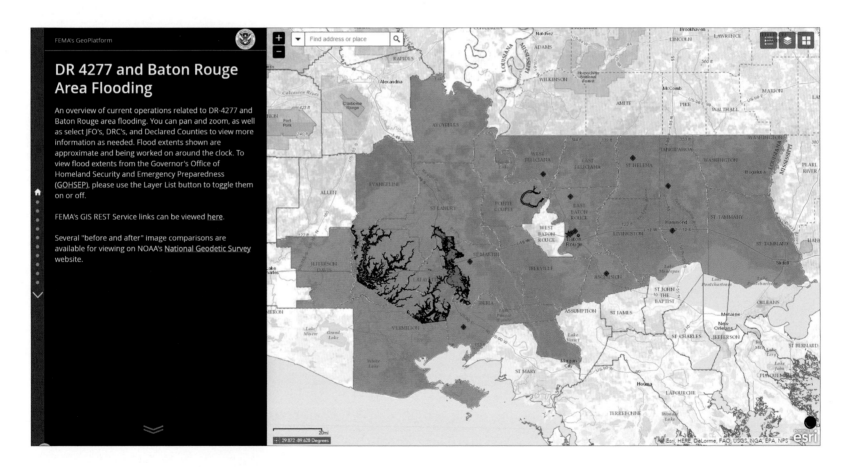

## Photo Survey

To gain additional situational awareness information from the "crowd" on where flooding had occurred, the Photo Survey application was configured to display hand-held photos taken by the Civil Air Patrol and have the public submit details on what was observed in the photos such as whether damages were visible, buildings were damaged, or roads were passable.

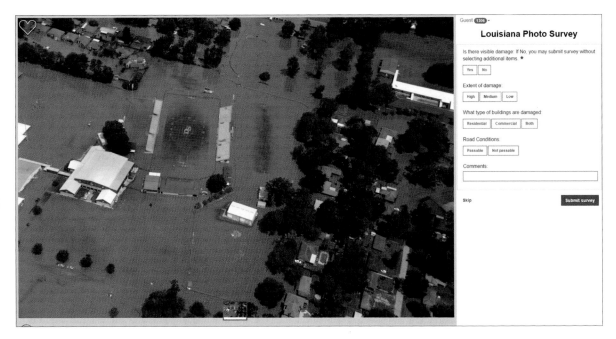

## FEMA Story Map Journal

The Federal Emergency Management Agency (FEMA) followed its standard protocol and created a "Disaster Journal" using an Esri Story Map template to convey information on current operations for the flooding response and recovery. This product contained information on the depth of flooding, damage estimates, current transportation status, disaster assistance, and much more.

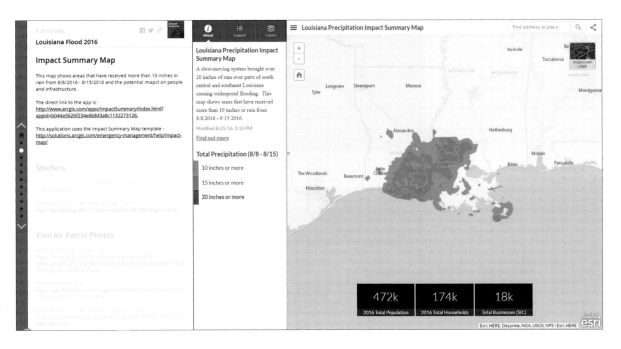

# Tropical Cyclone Exposure in the United States

The Baldwin Group/National Oceanic and Atmospheric Administration, Office for Coastal Management

Using data developed by MarineCadastre.gov, this story map highlights the modeled, historical exposure of the United States' offshore and coastal waters to tropical cyclone activity for the period 1900–2013. This story map contributes to the marine planning community and the overall effort of MarineCadastre.gov. Planners will be able to use tropical cyclone exposure data to better understand the susceptibility of marine waters and offshore activities to damaging winds. This data can help planners understand how tropical cyclones can impair offshore infrastructure or interrupt commerce and marine operations.

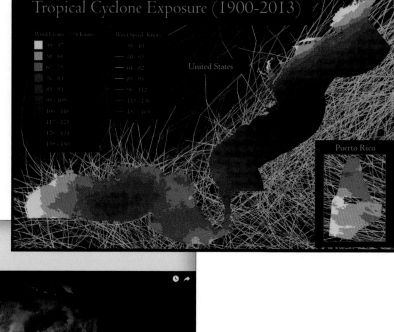

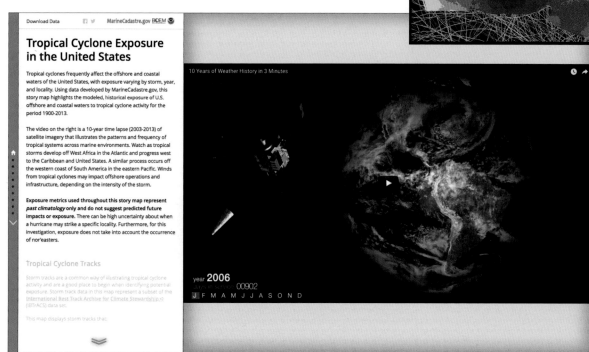

# Tsunami Information Service

National Oceanic and Atmospheric Administration, National Ocean Service

The National Oceanic and Atmospheric Administration Office for Coastal Management, the State of Hawaii, and Guam Civil Defense are using the Tsunami Information Service tool to address tsunami risks across the state and territory. The service provides evacuation maps and tsunami information in one central location to keep the public informed in Hawaii and Guam. This tool increases awareness of coastal hazards in any community and allows users to simply enter an address to identify hazard risks near their homes or workplaces.

## Tsunami Information Service

The Tsunami Zone Evacuation Tool increases awareness of coastal hazards in any community. The tool allows users to simply enter an address to identify hazard risks near their homes or workplaces. The tool provides an easy-to-understand hazard risk map, which is accompanied by related educational and awareness information, such as evacuation instructions or sheltering procedures.

The NOAA Office for Coastal Management, the State of Hawaii, and Guam Civil Defense are using the Tsunami Information Service tool to address tsunami risks across the state and territory. The Tsunami Information Service provides a one-stop shop for island residents and visitors to access tsunami-related information.

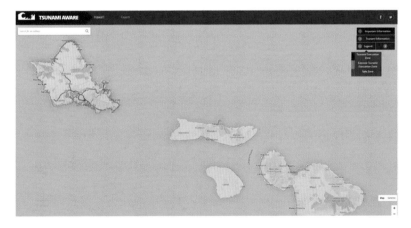

**NOAA Office for Coastal Mangement – Pacific Islands Office**

# Integrated Reporting of Wildfire Information

US Department of the Interior/Bureau of Indian Affairs/Bureau of Land Management/National Park Service/US Fish and Wildlife Service/US Department of Agriculture Forest Service/National Association of State Foresters/International Association of Fire Chiefs/Intertribal Timber Council/US Fire Administration

Integrated Reporting of Wildfire Information (IRWIN) is a system that serves federal and nonfederal partners that make up the wildland fire community. The IRWIN team is tasked with providing data exchange capabilities between existing systems used to manage data related to wildland fire incidents. IRWIN uses agile project management principles to build the data integration service for use in wildland fire management and operations. IRWIN orchestrates data based on business-defined requirements, using data standards defined by the National Wildfire Coordinating Group (NWCG), and authoritative data sources identified by the wildland fire community. This capability uses ArcGIS Server to orchestrate the data exchange between systems, validate that the data meets system requirements and is being passed from authoritative data sources.

Fire information such as location, size, and resources is often repeatedly entered into stand-alone systems as a foundation for their capabilities. More timely and accurate information must be entered into systems as conditions change over the life of an incident, while the original, outdated data remains in the supporting systems. An example is the location of a fire (latitude and longitude). An efficiency study identified that a dispatcher may enter this piece of data up to twenty-six times into different systems. Once dispatchers received the required information from each system, they didn't go back and update each system as more up-to-date location information became available. When questions arose about individual fires, there were multiple answers depending on which data source was queried for the answer. Although all the answers may be valid in their specific context, there was no authoritative data source for a consistent answer. This presented a challenge for the community and managers at all levels of the wildland fire community.

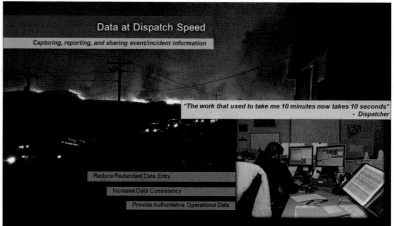

Today, IRWIN provides the capability for the dispatcher to enter data about a reported wildfire in one system. Then that system submits the data to the IRWIN data integration service and IRWIN makes that data available to other systems. IRWIN's success is that it reduces redundant data entry, increases data consistency, and provides authoritative operational data in near real time. During a recent dispatch training session, when asked what the dispatchers liked about IRWIN, 60 percent of them said that the work that used to take 10 minutes now takes 10 seconds. It not only reduced data input time but it also improved quality of work life in an inherently stressful job. The wildfire dispatcher is now able to focus on dispatching resources and providing support to wildland firefighters.

Upon deployment of the first version of IRWIN and during the wildland fire season, the wildfire community requested a view into the status and transaction history of the incident data flowing through the data integration service. It was quickly realized that the data flowing through the integration service may be the most robust data the community could access to support management decisions.

Using Portal for ArcGIS, the IRWIN team built the IRWIN Observer as a tool to meet the needs of the community. The IRWIN Observer allows the community not only a view of the data, but also provides access to the data. This data is used to perform analytics-based issue management decisions for operational and planning purposes. The IRWIN Observer allows decision-makers to analyze the data, make informed risk-based decisions, and report on the current situation and planned actions. The data is delivered in near real time showing transactional history for each incident over time. The viewer promotes complete transparency into the data integration throughout the enterprise.

The IRWIN Observer also serves as a viewer to monitor the status of the systems that are connected to the data integration service. At this time there are six systems creating, reading, and updating incident data through the IRWIN data integration service. There are twelve systems that are only reading data from IRWIN and over the next few years there are a number of federal and nonfederal systems scheduled to connect to the IRWIN data integration service. As these systems continue to connect to the service, and spatial coverage and completeness of the fire records improve, the wildfire community will be able to confidently tell the story of wildland fire management successes to all levels of the organizations, Congress, and the public.

# Flood Inundation Mapping for the City of Boise

Office of US Senator James E. Risch

Through maps on Senator James E. Risch's website, his constituents can visualize such important geographical information as locations of active forest fires within the state, places of congressionally funded projects, and flooding risk. Using National Weather Service data, a flood inundation map for the city of Boise displayed in ArcGIS visualizes the impact on the city center if significant spring rain and snowmelt runoff occurs.

# Flash Flood Potential Index

National Oceanic and Atmospheric Administration, National Weather Service

The National Weather Service issues flash flood warnings that protect life and property. This map gives meteorologists another perspective on the northern Mid-Atlantic region. The landscape of this region is diverse, and combining slope, soil type/texture, and land cover can address areas that cause more runoff than others. Urban areas are more prone to flash flooding but some rural areas are also at risk. The Flash Flood Potential Index shows river basins that are more or less prone to flash flooding.

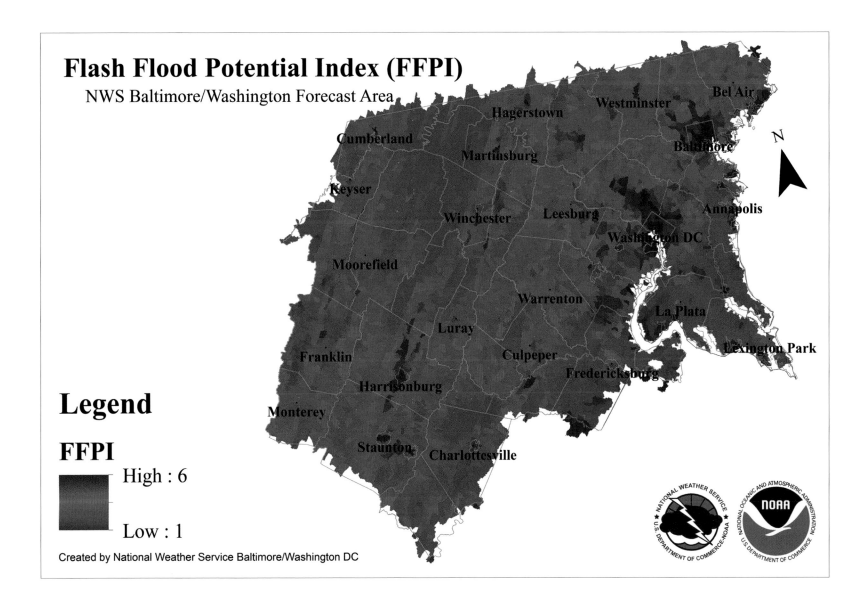

# The United States Military Academy

US Army

This map was produced for the United States Military Academy Association of Graduates, the institution's alumni organization, for its Register of Graduates book to update an older map. The updated map of the West Point area included ongoing new construction projects. This map provides the map viewer with a clear visualization of the terrain, vegetation, and buildings at West Point.

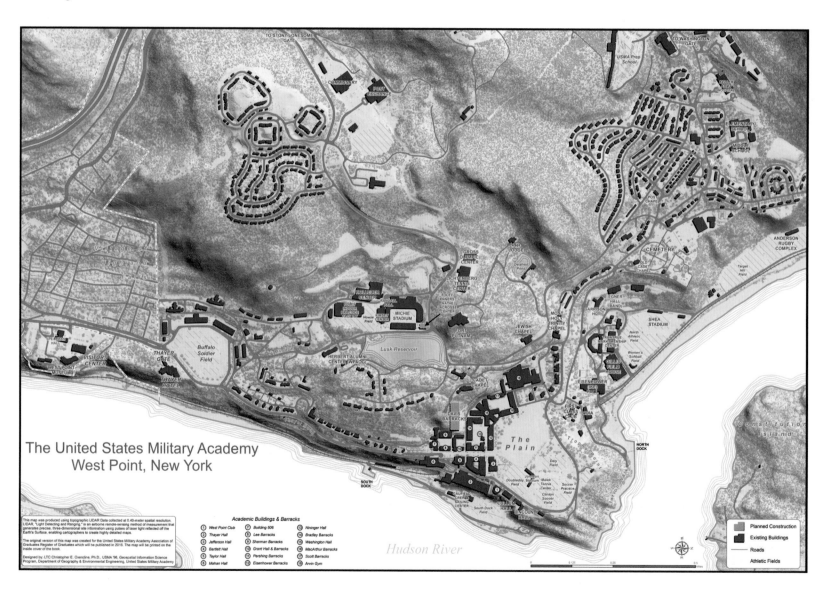

# Evaluating the Mysteries of Seismicity in Oklahoma

US Department of Energy, National Energy Technology Laboratory

This story map evaluates the factors related to the increase in seismicity in Oklahoma since 2008. The web app provides an educational tool for researchers to understand the complexities of the increase. It provides a better understanding of the various earthquake factors and data involved in order to evaluate the causes and risks for future seismic events in Oklahoma.

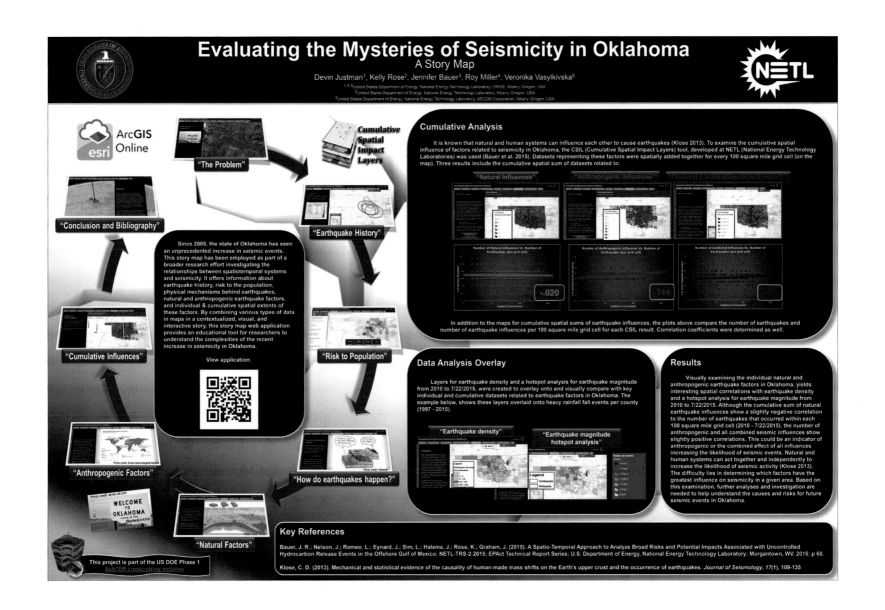

# Attacks on US Federal Government

US Secret Service, National Threat Assessment Center

This map displays geocoded locations of attacks against the US federal government that occurred between 2001 and 2013, exhibiting the type of weapon used in each attack. The map facilitates information sharing with stakeholders about the location and prevalence of attacks against the federal government.

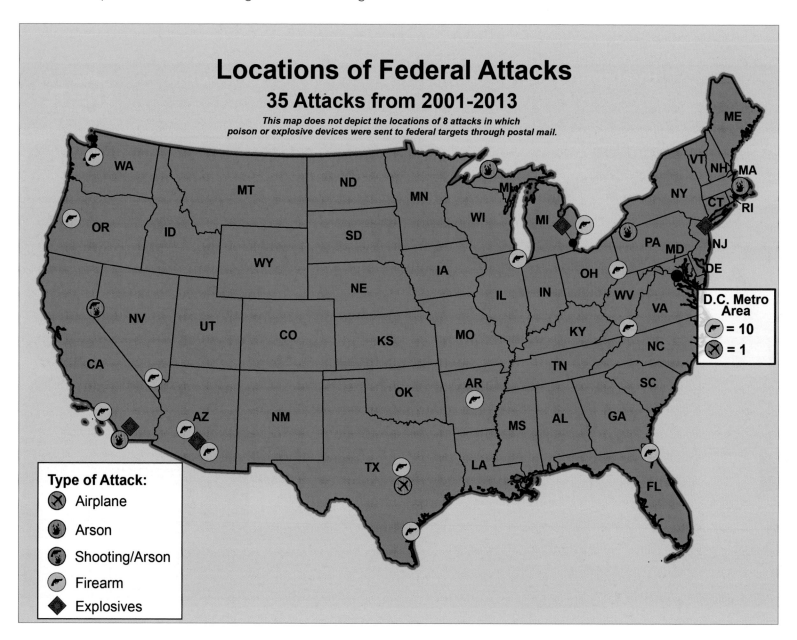

**Locations of Federal Attacks**

**35 Attacks from 2001-2013**

*This map does not depict the locations of 8 attacks in which poison or explosive devices were sent to federal targets through postal mail.*

**D.C. Metro Area**
- 🔫 = 10
- ⊗ = 1

**Type of Attack:**
- ⊗ Airplane
- Arson
- Shooting/Arson
- 🔫 Firearm
- ◆ Explosives

# Persons Traveling into the United States at Land Border Crossings and International Airports (2014)

US Department of Transportation

This map uses proportional symbols to show the number of persons that crossed into the United States in 2014 at land border crossings and international airports. The map supported a report by the US Department of Transportation's Bureau of Transportation Statistics on passenger travel in the United States.

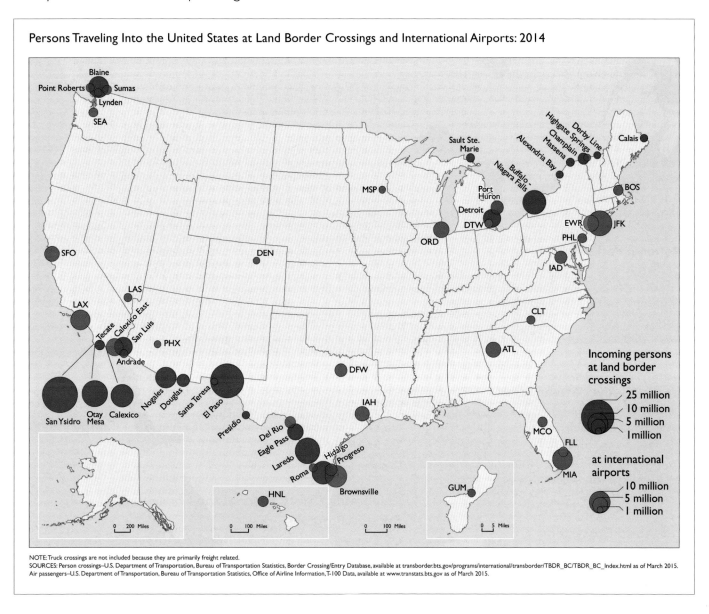

Persons Traveling Into the United States at Land Border Crossings and International Airports: 2014

Incoming persons at land border crossings
- 25 million
- 10 million
- 5 million
- 1 million

at international airports
- 10 million
- 5 million
- 1 million

NOTE: Truck crossings are not included because they are primarily freight related.
SOURCES: Person crossings–U.S. Department of Transportation, Bureau of Transportation Statistics, Border Crossing/Entry Database, available at transborder.bts.gov/programs/international/transborder/TBDR_BC/TBDR_BC_Index.html as of March 2015.
Air passengers–U.S. Department of Transportation, Bureau of Transportation Statistics, Office of Airline Information, T-100 Data, available at www.transtats.bts.gov as of March 2015.

# Credits

## Environmental Protection and Conservation

**Keep the Canyon Grand**
**Grand Canyon Trust**
SOURCES: National Park Service, Esri, HERE, DeLorme, Food and Agriculture Organization of the United Nations, US Geological Survey, US Environmental Protection Agency, Uranium Claims, Environmental Working Group; Mature & Old Growth Forest, US Department of Agriculture Forest Service, Wildlife Movement, Grand Canyon Wildlands Council

**100 Years of the National Park Service**
**National Park Service (NPS)**
SOURCES: NPS, NPMap, MapBox, Earthstar Geographics CNES/Airbus OS, US Department of Agriculture Farm Services Agency, Digital Globe, Microsoft

**Explore Virginia's National Parks**
**Office of US Senator Tim Kaine**
SOURCES: National Park Service, Esri, DeLorme, US Geological Survey, National Geospatial-Intelligence Agency, US Environmental Protection Agency, US Department of Agriculture

**Maps from Senate Analytical Mapping System: Arrowrock Dam Height Analysis, Boise River Streamflow Discharge Analysis)**
**Office of US Senator James E. Risch**
SOURCES: Esri, HERE, DeLorme, Earthstar Geographics, Food and Agriculture Organization of the United Nations, National Oceanic and Atmospheric Administration, US Geological Survey

**Explore a Tapestry of World Ecosystems**
**US Geological Survey (USGS)**
SOURCES: USGS, Esri, Delorme, NaturalVue, General Bathymetric Chart of the Oceans

**Coastal Community Vulnerability Assessment in Talbot County, Maryland**
**National Oceanic and Atmospheric Administration (NOAA), National Centers for Coastal Ocean Science**
SOURCE: NOAA, National Centers for Coastal Ocean Science

**Smart Location Calculator**
**US General Services Administration (GSA)**
SOURCES: GSA, US Environmental Protection Agency

**Monarch Butterfly Conservation**
**US Fish and Wildlife Service (USFWS)**
SOURCE: USFWS, Service Branch of Habitat Restoration

**Steens Mountain Wilderness Area**
**US Bureau of Land Management (BLM), Oregon State Office**
SOURCE: US BLM, Oregon State Office

**Lewis and Clark's Scientific Discoveries: Animals**
**National Park Service (NPS)**
SOURCES: NPS, Esri

**US Department of Agriculture CarbonScapes**
**US Department of Agriculture (USDA)**
SOURCE: USDA

**Methods for Locating Legacy Wells in America's Oldest Oil Field: Oil Creek State Park**
**US Department of Energy (DOE) National Energy Technology Laboratory (NETL)**
SOURCE: US DOE, NETL

**Celebrating Thirty Years of the Conservation Reserve Program**
**US Department of Agriculture, Farm Service Agency**
SOURCE: USDA Conservation Reserve Program

**Visualize Your Water Challenge for High School Students**

**Understanding Eutrophication in the Chesapeake Bay**
Washington-Lee High School/Nicholas Oliveira
SOURCES: US Department of Agriculture, Esri, HERE, DeLorme, National Geospatial-Intelligence Agency, US Geological Survey, National Park Service

**The Bonds of Water**
Poolesville High School/Sam Hull
SOURCES: Earthstar Geographics, David Rumsey Historical Map Collection, Esri, HERE, DeLorme, National Geospatial-Intelligence Agency, US Geological Survey, National Park Service

**Nutrient Pollution, the Bay's Biggest Threat**
Poolesville High School/Alex Jin
SOURCES: Earthstar Geographics, Chesapeake Bay Foundation, National Oceanic and Atmospheric Administration, National Ocean Service; World Resources Institute

**The Chesapeake Bay - A National Treasure in Trouble**
Poolesville High School/Clara Benadon
SOURCES: Esri, General Bathymetric Chart of the Oceans, DeLorme, NaturalVue, International Hydrographic Organization/Intergovernmental Oceanographic Commission, National Geodetic Survey

**Eutrophication in the Chesapeake Bay: Fertilizer and Manure**
Washington-Lee High School/Anna Lujan
SOURCES: Esri, General Bathymetric Chart of the Oceans, DeLorme, NaturalVue, International Hydrographic Organization/Intergovernmental Oceanographic Commission, National Geodetic Survey

# Socioeconomics

**Zika Maps**
Centers for Disease Control and Prevention (CDC), Food and Drug Administration (FDA), National Institutes of Health (NIH)
SOURCES: CDC, FDA, NIH, Earthstar Geographics, Esri, HERE, DeLorme

**Integrated Biosurveillance: Alert and Response Operations Products**
Armed Forces Health Surveillance Branch (AFHSB)
SOURCE: AFHSB

**Veteran Homelessness**
Office of US Congressman Mark Takano
SOURCES: Esri, HERE, DeLorme, National Geospatial-Intelligence Agency, US Geological Survey

**Our Nation's Veterans**
US Census Bureau
SOURCE: US Census Bureau's 2013 American Community Survey 5-Year Estimates

**Second Amendment Sentiment**
Office of US Senator James E. Risch
SOURCES: Esri, HERE, DeLorme, Earthstar Geographics, Food and Agriculture Organization of the United Nations, National Oceanic and Atmospheric Administration, US Geological Survey

**Percentage of Medicare Beneficiaries Enrolled in Medicare Advantage and TRICARE for Life by County**
Congressional Research Service, Library of Congress
SOURCES: US Census Bureau, the Centers for Medicare and Medicaid Services, US Department of Defense, Esri

# Credits (continued)

**Multi-Resolution Data Store Automated Map Output**
Ordnance Survey Ireland (OSi)
SOURCE: OSi

**Visualizing Which Provider to Pursue in a Case of Health-Care Provider Fraud**
US Postal Service (USPS), Office of Inspector General (OIG)
SOURCE: USPS, OIG

**Visualizing Where Mail Is Delayed in Processing Plants**
US Postal Service (USPS), Office of Inspector General (OIG)
SOURCES: USPS, OIG

**Social Security's Old-Age, Survivors, and Disability Insurance Beneficiaries and Benefits by State (2014)**
Social Security Administration (SSA), Office of the Chief Strategic Officer
SOURCES: SSA, Esri, HERE, DeLorme, Food and Agriculture Organization of the United Nations, National Oceanic and Atmospheric Administration, US Geological Survey, US Environmental Protection Agency

**Counties with No Opioid Treatment Program Facilities, by Level of Urbanization (2014)**
White House Office of National Drug Control Policy
SOURCES: Substance Abuse and Mental Health Services Administration Survey of Substance Abuse Treatment Services, National Center for Health Statistics

**Drug Overdose Deaths Involving Heroin and Other Opioids, by County (2010–2014)**
White House Office of National Drug Control Policy
SOURCE: US Centers for Disease Control and Prevention

**Laws Increasing Access to Naloxone, a Life-Saving Overdose Reversal Drug**
White House Office of National Drug Control Policy
SOURCE: The Network for Public Health Law

**Medicaid Expenditures on Sovaldi and Patients Treated (2014)**
Office of US Senator Ron Wyden
SOURCES: State Medicaid programs

**Bullseye Mapping Tool**
US General Services Administration (GSA)
SOURCES: US GSA, REIS, CBRE, Federal Emergency Management Agency, Core Based Statistical Areas

**General Services Administration Fiscal Year 2017 Budget Request**
US General Services Administration (GSA)
SOURCES: US GSA, City of Indianapolis, Esri, HERE, DeLorme, INCREMENT P, Intermap, US Geological Survey, US Environmental Protection Agency, US Department of Agriculture

**Voter Turnout in Chicago (2008–2015)**
University of Chicago
SOURCE: Chicago Board of Elections

**Averaged Eligibility Map**
US Department of Agriculture (USDA), Food and Nutrition Service
SOURCES: USDA, Esri, HERE, DeLorme, National Geospatial-Intelligence Agency, Share Our Strength, US Census, IncrementP, City of Olympia

**Highly Pathogenic Avian Influenza Scenarios Affecting Poultry**
US Department of Agriculture (USDA),
SOURCES: USDA Emergency Programs Division, Office of Homeland Security and Emergency Coordination; USDA Animal and Plant Health Inspection Service

**Crop Migration and Change for Corn, Soybeans, and Spring Wheat in the North Central United States (2006–2015)**
US Department of Agriculture (USDA) National Agricultural Statistics Service (NASS)
SOURCE: USDA, NASS

**Using GIS for the Albanian Census**
Albanian Institute of Statistics (INSTAT)
SOURCE: INSTAT

**Geocube: A Flexible Energy and Natural Systems Web Mapping Tool**
Mid-Atlantic Technology, Research, and Innovation Center (MATRIC)/US Department of Energy (DOE), National Energy Technology Laboratory (NETL)
SOURCES: US DOE, NETL; Bureau of Ocean Energy Management; Minerals Management Service; US Fish and Wildlife Service; National Oceanic and Atmospheric Administration

**STATcompiler: The DHS Program**
Blue Raster/US Agency for International Development (USAID)
SOURCE: Demographic and Health Surveys Program of USAID

**Water More Precious Than Gold: The Story of the Boise Project**
US Bureau of Reclamation, Pacific Northwest Region
SOURCE: US Bureau of Reclamation, Pacific Northwest Region

**MapEd**
Blue Raster/National Center for Educational Statistics (NCES)
SOURCES: NCES, US Census Bureau

**The Hidden Cost of Suspension**
Blue Raster/US Department of Education (ED)
SOURCES: US ED, Esri, HERE, DeLorme, MapmyIndia, OpenStreetMap contributors and the GIS user community, Civil Rights Data Collection 2011–2012

**Naval Station Norfolk Measurements of Energy Reduction**
GISInc.
SOURCE: Navy Shore Geospatial Energy Module

**Top Twenty-Five US–International Trade Freight Gateways by Value of Shipments (2014)**
US Department of Transportation (DOT)
SOURCES: US DOT, US Census Bureau, Foreign Trade Division; USA Trade Online; US Army Corps of Engineers, Navigation Data Center; US Department of Transportation, Bureau of Transportation Statistics; North American TransBorder Freight Data

**The Bartholdi Fountain**
House Historian's Office
SOURCE: Historian of the US House of Representatives, Library of Congress

# Credits (continued)

## National Security

**Louisiana Flood Maps**
**Federal Emergency Management Agency (FEMA)**
SOURCES: FEMA, Esri, HERE, DeLorme, UN Food and Agriculture Organization, US Geological Survey, National Geospatial-Intelligence Agency, US Environmental Protection Agency, National Park Service

**Tropical Cyclone Exposure in the United States**
**The Baldwin Group/National Oceanic and Atmospheric Administration (NOAA), Office for Coastal Management**
SOURCES: NOAA, Bureau of Ocean Energy Management, MarineCadastre.gov

**Tsunami Information Service**
**National Oceanic and Atmospheric Administration (NOAA), National Ocean Service (NOS)**
SOURCES: NOAA, NOS; Hawaii Emergency Management; Guam Civil Defense, City and County of Honolulu; Maui County; Kauai County

**Integrated Reporting of Wildfire Information**
**US Department of the Interior et al.**
SOURCES: US Department of the Interior, Bureau of Indian Affairs, Bureau of Land Management, National Park Service, US Fish and Wildlife Service, US Department of Agriculture Forest Service, National Association of State Foresters, International Association of Fire Chiefs, Intertribal Timber Council, and US Fire Administration

**Flood Inundation Mapping for the City of Boise**
**Office of US Senator James E. Risch**
SOURCES: Esri, HERE, DeLorme, Earthstar Geographics, Food and Agriculture Organization of the United Nations, National Oceanic and Atmospheric Administration, US Geological Survey

**Flash Flood Potential Index**
**National Oceanic and Atmospheric Administration (NOAA), National Weather Service (NWS)**
SOURCE: NOAA, NWS

**The United States Military Academy**
**US Army**
SOURCE: United States Military Academy

**Evaluating the Mysteries of Seismicity in Oklahoma**
**US Department of Energy (DOE), National Energy Technology Laboratory (NETL)**
SOURCE: DOE, NETL

**Attacks on US Federal Government**
**US Secret Service, National Threat Assessment Center**
SOURCE: US Secret Service's National Threat Assessment Center

**Persons Traveling into the United States at Land Border Crossings and International Airports (2014)**
**US Department of Transportation (DOT)**
SOURCES: US DOT, Bureau of Transportation Statistics, Border Crossing/Entry Database; Office of Airline Information; T100 Data